Praise for

Letters from the

"What may be the single best book on bees." —*Booklist*

"[An] elegant book, modest and evenly paced, that's dense-packed with information, but such a pleasure to read that you hardly notice how much you're assimilating, page by page." —Salon

"*Letters from the Hive* is the engrossing story of our long and richly layered relationship with bees. It reminds us of the fragile interconnections between all the creatures on this earth."

—Alice Waters, owner, Chez Panisse restaurant

"Beekeeper and entomologist Buchmann brings together scientific rigor and environmental zeal in a passionate history of the relationships among people, honey, and bees." —*Kirkus Reviews*

"The religion of the bee, the art and the biology of beekeeping, one man's love and his race's long fascination with the honey bee, it's all here—up to and including the treatment of cataracts, how to avoid getting stung, and a recipe for lasagne with honey. *Letters from the Hive* provides everything a beekeeper's apprentice could ask for."

—Laurie R. King, author of *The Beekeeper's Apprentice*

"*Letters from the Hive* is really an extended love letter—a charming, enthralling, and deeply authoritative window into the sweet, age-old affair between humans and the honey makers. From the seasons of the hive to the history, properties, and diversity of honey, Buchmann brings us into the confidence of the bees as only he could possibly do."

—Robert Michael Pyle, author of
Chasing Monarchs: Migrating with the Butterflies of Passage

"Like a gracious host . . . Buchmann introduces bees to the reader with colorful vignettes, bits of folklore, [and] recipes for cooking with honey. . . . He also writes as an activist—one who has done extensive research into the beekeeping of the Mayans, whose tradition he documents in hopes of preserving it before it dies out."

—*Christian Science Monitor*

"Stephen Buchmann has done more to advance the conservation of all bees—and the flowers that depend upon them—than any other human in history. This intimate history is one very creative soul's lifework. Savor it."

—Gary Nabhan, author of *Cross-Pollinations: The Marriage of Science and Poetry*

"Stephen Buchmann's *Letters from the Hive* is a fascinating and lovingly informative account of one of humanity's greatest accomplishments, our symbiosis with the honey bees."

—Edward O. Wilson, Pulitzer Prize–winning author of *On Human Nature* and professor emeritus at Harvard University

Letters from the Hive

An Intimate History of Bees, Honey, and Humankind

Stephen Buchmann
with Banning Repplier

BANTAM BOOKS

LETTERS FROM THE HIVE
A Bantam Book

PUBLISHING HISTORY
Bantam hardcover edition published May 2005
Bantam trade paperback edition / June 2006

Published by Bantam Dell
A Division of Random House, Inc.
New York, New York

All rights reserved
Copyright © 2005 by Stephen Buchmann and Banning Repplier

Cover art by Emile Claus/The Bridgeman Art Library
Cover design by Belina Huey

Book design by Virginia Norey

Library of Congress Catalog Card Number: 2004065580

Bantam Books and the rooster colophon are
registered trademarks of Random House, Inc.

ISBN-13: 978-0-553-38266-2
ISBN-10: 0-553-38266-7

Printed in the United States of America
Published simultaneously in Canada

www.bantamdell.com

BVG 10 9 8 7 6 5 4 3 2 1

For my daughters, Marlyse and Melissa,
remembering shared bee adventures in our Tucson backyard
and across the Sonoran desert, with love and gratitude
for making my life complete

Melissa means "bee" in Greek.
In Greek mythology, Melissa was a bee nymph,
the daughter of King Melisseus, who cared for the young Zeus.

Table of Contents

A Linguistic Note:

Honey Bee Versus Honeybee

Throughout this book, we have used the term *honey bee* rather than the dictionary-approved *honeybee*. The late insect morphologist Robert E. Snodgrass, in his classic 1956 work on the anatomy of the honey bee, made a very cogent argument for the use of the two-word name over the compound form:

> *Regardless of dictionaries, we have in entomology a rule for insect common names that can be followed. It says: If the insect is what the name implies, write the two words separately; otherwise run them together. Thus we have such names as house fly, blow fly, and robber fly contrasted with dragonfly, caddicefly, and butterfly, because the latter are not flies, just as an aphislion is not a lion and a silverfish is not a fish. The honey bee is an insect and is preeminently a bee; "honeybee" is equivalent to "Johnsmith."*

Letters from the Hive

Introduction

The more we know of other forms of life, the more we enjoy and respect ourselves. . . . Humanity is exalted not because we are so far above other living creatures, but because knowing them well elevates the very concept of life.

—Edward O. Wilson,
Biophilia: The Human Bond with Other Species

Often, in quiet moments, I like to contemplate the path that brought me to where I am today. The curious little boy who kept pet fence lizards, raised caterpillars, chased butterflies, chipped fossils out of rocks, and read every nature guide available has metamorphosed into a man with the same deep connection to nature—and a profession that is a direct outgrowth of those early interests. But with the passage of years my childhood passion has matured; I now have a sense of the fragility of the natural world I love so much, and I have dedicated my life to doing whatever I can to help preserve it.

As an entomologist specializing in plant pollination by bees, my research has taken me to Australia, Belize, Brazil, Costa Rica, England, Germany, Israel, Malaysia, Mexico, Panama, South Africa, Thailand, and nearly every state in

the union. Everywhere I've been, the story is the same—the once vast wilderness, from spectacular desert landscapes to lush, steaming rainforests, has been chopped up and reduced to isolated islands. Unfortunately, we think only about ourselves, and having inscribed our tale upon the earth, we have little concern for the flora and fauna that share our very finite planet.

About a dozen years ago, I began to think long and hard about the academic research articles I was writing. Was anyone out there listening to what I was saying, did they care, was my work relevant? Was it enough simply to produce well-honed scientific papers and publish them in prestigious journals read only by my colleagues? I had a terrible suspicion that in fact it was not nearly enough and that I should be doing a great deal more to alert the world to the catastrophe it was facing.

It was at that time, with so much of the beauty and diversity of my beloved planet disappearing—under the plow, the parking lot, the subdivision, and the mall—that I began to change my focus. I became a conservationist, one with an evangelical fervor to spread the word and help stop, or at least slow down, the destruction.

Much of the research I now do is directed at helping to investigate and solve the problem of our disappearing pollinators—the bees and other insects that, by spreading pollen from one plant to another, enable them to reproduce. As you'll see in Chapter 5, the world would be a desolate place without our pollinating friends, devoid of flowers, flowering plants, and many of our favorite foods, including honey—the one I'll be celebrating later in this book.

Being an amateur beekeeper (as well as a professional naturalist, researcher, and conservationist) is deeply important to me. I've kept honey bee colonies intermittently since my very first one back in 1970. To me, beekeeping is about staying in touch with the pulse of the planet, with its rainforests, deserts, wetlands, and meadows. Living with bees and investigating their ways gives me vital insights into the cycles of nature that enable the ecological world to function. Bees fly out and gather information about the flora in their part of the world, then bring that data back to me. I, in turn, gather information for my work by studying the honey that results from those floral visitations. Thanks to bee watching and honey analysis, I am acutely attuned to my environment—and to the dangers that threaten it.

Making my living from bees and flowers is about much more than getting a paycheck. The more than twenty thousand different kinds of bees found worldwide are my touchstone, allowing me to escape the stressful human environment of concrete, steel, and glass. After hours spent in the confines of a windowless office dominated by a glaring computer screen, my aching head and sore neck muscles tell me it's time for a walk in the desert to see what the bees are up to. Nature has the power to teach, to renew, and to heal, something we must never forget.

Stung by the Biophilia Bee

When I was growing up in Orange County in Southern California, my family lived close to Tonner Canyon and the sights, sounds, and smells of the chaparral. Southern California in the late 1950s and early 1960s was much less crowded and polluted than it is today. There were still plenty of wild areas where the dramas of natural history could unfold before my eyes—praying mantids grabbing a quick meal, crab spiders prowling in flowers for hidden treasure, caterpillars spinning their cocoons, scurrying salamanders and newts, shy tadpoles and froglets. By the time I was eight years old, I had started to collect insects, filling countless jars in my bedroom museum with the latest bug du jour. I owe a debt of gratitude to my late mother for allowing me to keep this Lilliputian livestock indoors. She didn't always encourage my insect studies, but she didn't forbid them either. During high school, my fascination with bees and flowers blossomed. It all began with a rather misguided (and painfully inept) attempt to plunder a bee nest for honey. Despite well-stung ankles that were so swollen they looked like footballs, and itching that went on long after the pain had subsided, my interest in bees continued, and it only intensified when I went away to college. I'm sure my parents wondered when, if ever, I was going to grow out of my bug phase. I'm thankful I never did. I'm still the ever-curious kid, the insatiable backyard naturalist who has been lucky enough never to grow up. Instead, I've followed my childhood dreams with childlike abandon all the way into adulthood.

Now I've got what could be called a bad case of the biophilia bug. *Biophilia* is a term coined by Harvard entomologist and two-time Pulitzer Prize winner Edward O. Wilson. Essentially, it means that all humans, whether we are aware of it or not, have an innate need for contact with a diversity of other life-forms, not just our own. We feel assured, comfortable, and happy when we know that there are other animals about. This explains the popularity of zoos and public aquariums, which receive more visits per year than all sporting events combined. We're also a nation of sixty million backyard gardeners, indulging a passion that keeps us in touch with the rhythms of nature. Most of us prefer vistas that include water, verdant meadows, and forests to sterile cityscapes, barren of life but for the onrush of humans and cars. And for those who live in cities, views of water such as the one from Lake Shore Drive in Chicago, or of an open, green expanse, like that seen from the buildings along Fifth Avenue and Central Park West in New York, command the highest real estate prices. Even residents of apartments with no views at all reach out for nature, if only in the form of paintings, photographs, or a window box filled with potted plants. Some urban dwellers are more ambitious, of course. There are actually beekeepers who tend their hives on New York and Parisian rooftops.

Biophilia, an idea that has fired the imaginations of biologists and sociobiologists around the world, provides a powerful framework for understanding our proper place in nature. The premise of biophilia is the existence, in all of us, of genetically based physiological and neural structures that respond to differing habitats and various animals and plants in selective ways. Biophilia formalizes a human experience that goes all the way back to prehistoric times—namely, our deep, sustaining, almost sacramental relationship with the natural world. Simply put, *biophilia* means the love of life—all life.

The flip side of biophilia is biophobia. Around the world, but especially in the West, some people harbor deeply rooted fears of the wilderness. Feelings of revulsion toward cockroaches, spiders, snakes, and rats are commonplace. Biophobia can set a dangerous precedent, justifying the politics of exploitation, domination, and extermination of other life-forms.

But biophilia is the general rule. And birds and the so-called charismatic megafauna (giant pandas, elephants, and other large mammals) are particularly beloved. Honey bees, though tiny and equipped with stingers, are also loved and admired by millions for their industrious ways and sweet gift of honey. As you will see later in this book, bees and honey have long inspired the human imagination, playing an important role in the evolution of myths, religious rituals, social traditions, and even a few taboos. People have been robbing bees of their honey for at least as long as they have been leaving records of their exploits in the form of cave paintings, many of which document this risky enterprise. And the treasure they found has been used not just as food but in medicinal preparations and the distillation of delightful intoxicants.

As for me, bees are my life. I live for seeing, studying, and photographing bees of all kinds as well as the flowers they visit and pollinate. My obsession runs deep and includes tasting and cooking with the honey that bees produce. Bees inspire in me none of the dread or fear that many people feel for insects, but rather the kind of affection felt for engaging little animals. Seen under a hand lens or microscope, they endear themselves to us with their big compound eyes, long, sensitive antennae, and fuzzy brown and black striped abdomens. They not only make wonderful research organisms for scientists, they also make great miniature pets. Think about it. A colony of honey bees is a lot easier to care for than most domesticated creatures. In fact, they are very self-sufficient, requiring only a cavity to nest in and plenty of flowering, nectar-rich plants nearby.

Though many of us love bees almost as much as we love lions, giraffes, pandas, and elephants, we are tragically failing them by thoughtlessly destroying their habitats—the foraging areas, or "bee pastures," where they make their living grazing on flowers in their constant search for nectar and pollen. In fact, increasing environmental degradation is diminishing the quality of all our lives as well as our emotional and spiritual well-being. Real experiences among plants and animals in the great outdoors are required for our mental and physical health. We don't need more reality television programs or nature documentaries. We need to get outside and experience nature firsthand. As noted by my friend the naturalist and author Robert Michael Pyle, the loss of direct, personal contact with wildlife creates cycles of disaffection, apathy, and irresponsibility toward our precious planet and its myriad forms of life.

Bees and flowers are as vital a part of the intricate web of life as we ourselves are. The question we must then ask is: Do we love life enough to save it? If we do, we need to transform the way we use the earth's finite endowment of land, water, air, and wildlife. Unfortunately, we are as good at denial as we are bad at thinking in the long term or on a united global scale. Perhaps the solution lies in the recovery of that sense of wonder and amazement we all experienced as children when first discovering the plants and animals that surround us. I hope that an intimate look at the enduring bond between bees and mankind, forged in the sweetness of the honey pot, will rekindle that sense of wonder, helping us to stay in touch with our biophilic instincts and renew our age-old covenant with the world in which we live.

> *I have set before you life and death,*
> *blessing and cursing: therefore choose life,*
> *that both thou and thy seed may live.*
>
> Deuteronomy 30:19

Chapter 1

The Beginning of an Enduring Passion: Prehistoric Honey Hunters

O bees, sweet bees! I said: that nearest field
Is shining white with fragrant immortelles.
Fly swiftly there and drain those honey wells.

—Helen Hunt Jackson,
"My Bees"

A few years ago, when I was still keeping honey bees in my Tucson backyard, I always found it particularly exciting to check my hives in late spring, after the blooms had begun and the honey had started to flow. What had my bees been up to? What would the honey crop taste like? If generous winter and early spring rains had tickled the sandy desert soils and brought them to life, there could have been an explosion of wildflowers. Even with normal precipitation, the old desert standbys—velvet mesquite, foothill, and littleleaf paloverde trees—would break out in riotous bloom, attracting hungry bees from far and wide. Tens of thousands of them would plunder the flowers and carry home the precious nectar, transforming it into golden honey—a sweet, fragrant crop that I was always eager to sample.

Whenever I cracked the lid of one of my hives, the bees would rush up toward the light to see what had disrupted the cozy tranquility and comforting darkness of their nest. Selecting a middle frame fat with honey, I would ease it up and out of the box while the bees, clinging to its surfaces, ran about chaotically, confused by this rude interruption of their smooth, efficient daytime routine. Only days before, worker bees would have sealed the crinkly, textured surface of the honeycomb with the virgin white beeswax they secrete as tiny scales and form into cell caps through the workings of their mandibles.

This was the moment I had been longing for throughout the long winter months—a perfect day in late April with the first honey crop of the year ready to taste, waiting for me beneath those brilliant white caps. Since I never wore clumsy bee gloves, I was able to thrust my right forefinger deep into the comb and drag it across the frame, rupturing more than a hundred cells and releasing the glistening honey, which would stream out in thick little rivulets with the bees in hot pursuit. Withdrawing my finger, I would savor my prize, for there is nothing in the world like the taste of warm, fresh honey straight from the comb.

I am not alone in my passion for honey-making bees and their honey. From prehistoric times to the present, we humans have felt a mysterious and enduring connection to these furry little creatures and the food they produce. We have endowed them with magical properties, attributed to them surprising healing and cleansing powers, and seen in them meaningful symbols representing some of our most profoundly held beliefs.

Our fascination with bees is deeply rooted in our collective consciousness. We see it in the cave paintings that our prehistoric ancestors left behind. We can read it in the rich, complex rituals and traditions that evolved to govern our relationship with these admirable insects. And we can still catch the reverberations of our instinctive connection to that part of the natural world every time a husband calls his wife "honey" or an excited child chases a buzzing bee through a bright summer afternoon. But its influence is much more far-reaching than you might imagine, extending not just to everyday moments of affection and play but to diverse cultures, religious beliefs, cuisines, and scientific study around the world. We can look for its roots in our history and, before that, our prehistory.

Thanks to petroglyphs, the spectacular painted records still visible on cave walls throughout Europe, Africa, Asia, and even Australia, we know our ancestors definitely had a sweet tooth, and we know that they indulged it by embarking on arduous and often dangerous honey hunts, armed with tools that enabled them to pillage bee nests with remarkable efficiency. We don't know why cave artists put so much effort into recording these often dramatic hunts. Perhaps the honey hunts signified something more profound than the simple harvesting of an ingredient to sweeten their days—something with deep religious or ceremonial meaning. Whatever the reason, vivid paintings chronicling those honey-hunting expeditions—beautifully stylized yet powerfully real—have been found on the ceilings and walls of hundreds of caves spanning the globe.

In her recent book *The Rock Art of Honey Hunters*, Dr. Eva Crane, the grande dame of honey bee researchers, has collected some of the most striking examples of the cave art chronicling these prehistoric hunts. As she has vividly documented, there are a number of common elements that recur throughout this pictorial world. The honeycombs are prominently drawn, generally with great exuberance and appearing much larger than they are in real life. Bees, with or without wings, are shown flying angrily about as their nests are pillaged by the daring hunters. The hunters themselves are usually depicted either standing at the foot of a tree or cliff that harbors a bee nest or climbing long rope ladders to reach their prize. And they are typically shown naked—although to modern beekeepers, the idea of raiding a colony without protective clothing seems foolhardy at best. The honey hunters portrayed in African cave art frequently wear penis sheaths and nothing else.

When looking at photographs of some of these paintings, I can't help thinking of the brilliant frescoes that transformed the walls of medieval and Renaissance churches into graphic lessons in religious and moral history. The cave painters, who created their art thousands of years before, using mineral pigments mixed with animal fats, may also have wanted to give their work a didactic or sacred meaning. There are several clues that support this supposition: the cave walls that served as their canvases enclosed spaces large enough to have comfortably accommodated crowds of people gathered to watch or participate in ceremonial rituals; the subjects of the paintings—hunting and fertility rites—are the kind that lend themselves to sacramental reenactment; and, in addition to the honey- and animal-hunting scenes, some of the paintings show women engaged in what seems to be sacred dancing, groups of archers led by a priest or shaman wearing a large headdress, and warriors engaged in hand-to-hand combat.

I'd love to know what lessons, if any, our forebears found embedded in the painted chronicles of their honey hunts. Did the rituals of the hunt serve to enlighten them or to give them spiritual guidance? Did they inspire these nomadic clans to strive for the same kind of efficient, productive social organization that the bees had so wonderfully evolved? And what role did honey play in their daily lives? Was it a key ingredient in their primitive cuisine, a medicine used to cure a number of ills, or simply eaten raw as a palate-pleasing, energy-supplying snack after a long, hard day hunting and gathering?

Despite the dedicated work of many archaeologists, we'll probably never know the answers to these questions. But based on hunting scenes found in caves separated by thousands of miles (and executed with uncanny similarity), we can safely say that plundering bee nests has been an important human activity for many millennia, all over the globe. In fact, there are places in the world today where the ancient rituals of those long-ago honey hunters are still practiced virtually unchanged. In the rainforests of Malaysia, the remote valleys of Nepal, and the vast Australian outback, honey-hunting clans set out on expeditions so similar to those depicted in prehistoric cave art that those paintings might well have served as their primers. We will actually travel to Malaysia to witness some of those rituals as they unfold today. But first, let's travel back in time, to a cave that contains one of the most vivid of the ancient honey-hunting paintings.

We are at La Araña—the Cave of the Spider. It is one of many caves and prehistoric shelters honeycombing the limestone mountains of what is now the province of Valencia, near the city of the same name, on the east coast of Spain. Six thousand years ago, when the paintings at La Araña were executed, Europe had just entered the First Neolithic Age, which lasted for another thousand years.

Though we tend to think of the people who used these caves as cave dwellers, they were actually nomadic hunters and gatherers who lived in extended families or small bands and stayed in the rock shelters only for brief periods as they traveled with the seasons. The caves were often located in hillsides that offered wide views of animal migration routes and, as a result, prime hunting opportunities. The caves also provided protection from the elements and probably from natural predators and human enemies as well. Beyond their role as temporary sanctuaries, they served as burial chambers and meeting places for clans scattered over hundreds of miles. Thus the caves helped facilitate communication and intermarriage among these dispersed peoples.

The La Araña cave, explored in 1924 by Spanish archaeologist Hernández Pacheco, is set in a landscape of wild natural beauty, with rugged slopes, deep gullies, and clear-running rivers. Blackened soot on the low ceiling near the entrance indicates that early humans took shelter there, lighting fires to provide warmth and cook their meals, and recording the events of daily life in paintings on the interior walls.

Let's imagine a cold, damp winter afternoon six thousand years ago in the cave of La Araña. A small band of men, women, and children has gathered around a roaring blaze at the mouth of the cave to recount their day's activities, which include not just hunting and gathering but also crafting and maintaining a wide variety of tools, stitching animal skins together with bone needles to make warm clothing to fight off the bitter cold of winter, nursing infants, and preparing meals. Implements essential to their survival lay scattered about and include hand axes, cleavers, and scrapers to work hide and shape wood. Barbed harpoons and spears, made from bone and antler and decorated with carved animal designs as well as various ornamental pendants, can also be seen. The tools are well made, for this is a period of great innovation. Change is occurring at an increasingly rapid rate and will continue over the next centuries as agriculture takes root, the nomads settle in villages, a hunter-gatherer economy transitions to a producing one, and a whole new way of life is born. In fact, our friends in the cave may have already begun rudimentary farming to supplement the fruits of their hunting and gathering expeditions.

As the day wanes, dinner is served at La Araña. Items on the menu include reindeer, bison, ibex, horse, and red deer. And the pièce de résistance is honey, dripping thick and golden from the comb.

Like nearly everything our ancestors used, honey had to be pilfered from the natural world around them—right from the nests of the bees themselves. Except for a small number of tropical wasps and ants, no other creatures collect and store concentrated reserves of sugar the way honey bees do. In Spain, the honey would have come from the European honey bee *(Apis mellifera)*.

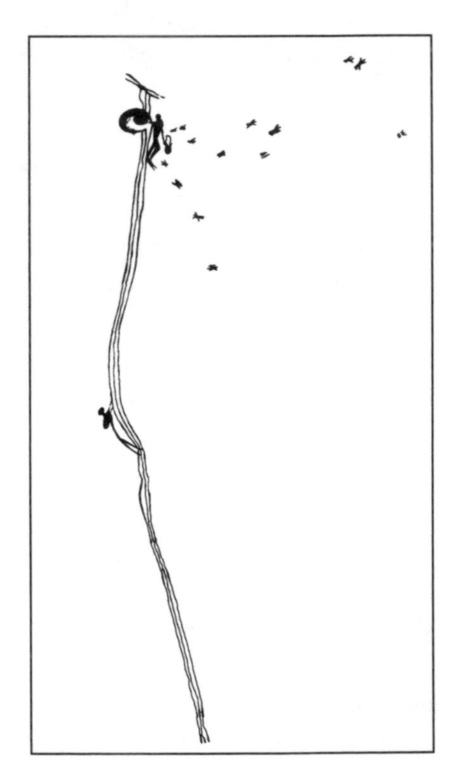

Petroglyph showing honey collection from a wild bee hive. Cave of the Spider, Valencia, Spain.

The challenges of harvesting honey were daunting, as we can see in the painting for which the Cave of the Spider is best known—a compelling rendition of the rigors and rituals of the honey hunt. Using a concavity in the rock wall to represent a bee nest, the painter drew a man climbing what appears to be a rope ladder and being attacked by a swarm of enormous bees defending their honey stores. The honey hunter is simply sketched, a fragile-looking stick figure, yet we sense skill, strength, and determination as he balances precariously on his flimsy ladder, enduring the painful stings and angry harassment of the bees. We also sense that something mysterious, something we can't quite grasp, is going on here. There is an almost spiritual quality to the painting, which suggests a tacit understanding between the bees and the barbarian who is attacking their stronghold. This is not simply a callous plundering of goods, a cold-blooded raid. It's as if we are witnessing a sacred contest, a battle of wills between equals. Perhaps the special relationship between humans and bees, which was to evolve into so many elaborate rituals and traditions over the millennia, had already begun to take shape.

The cave paintings leave no doubt that honey hunting has been going on for thousands of years. Long before the first humans descended from the thorny acacia trees of the African savannahs and began a new life as tool-making, fully bipedal primates, Malaysian honey bears, honey guide birds, and South African honey badgers were all plundering bee nests. But it was the ingenuity of our early ancestors that turned the honey hunt into a highly ritualized—and effective—activity. It probably didn't take long for prehistoric hunters and gatherers to discover that the nests of certain highly social bees—bumblebees *(Bombus)*, stingless bees *(Melipona* and *Trigona)*, and honey bees *(Apis mellifera* and other species in the genus *Apis)*—contained plentiful reserves of honey. And once they had figured that out, it was doubtless a short step to learning how to attack and exploit these tasty, energy-rich targets.

No one has been able to trace the evolution of the human sweet tooth, which is certainly at the root of our passion for bees and honey. We do know that our ape and chimpanzee relatives have a well-developed taste for sugar and aren't shy about availing themselves of any opportunity to gorge on it. And we also know that besides the honey from bees, the only other concentrated sources of sugar available to early humans would have been fruits, berries, and certain tropical grasses. So the honey stored by bees was a prize well worth enduring the stings delivered by the guardians of the nests. Other "spoils of war" from the nests included the juicy, protein-rich brood, consisting of bee larvae and pupae, along with equally nutritious pollen. It is little wonder that so many kinds of mammals and birds in the prehistoric world developed a taste for honey and young bees, along with the necessary skills for locating bee nests and breaking into their well-stocked pantries.

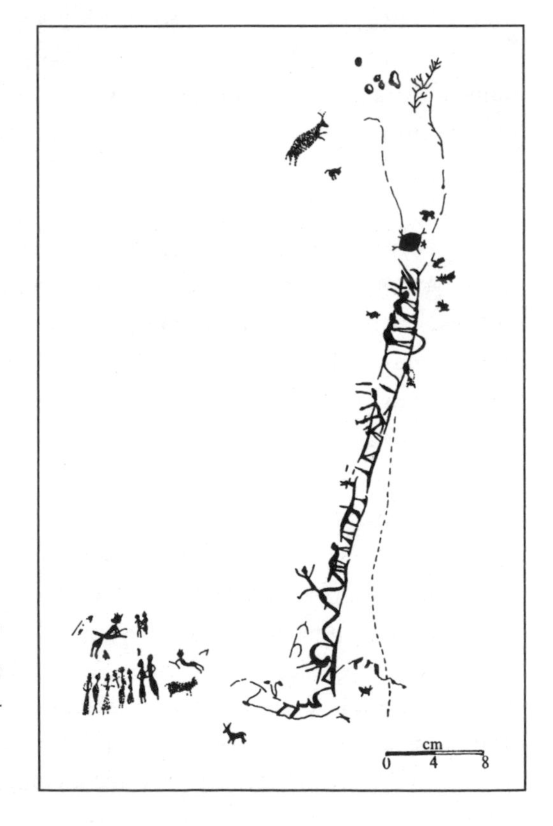

Petroglyph depicting honey hunters climbing a tall ladder to plunder a nest. Barranc Fondo, Castellón province, Spain.

La Araña is just one of many caves in Spain where we can see paintings that memorialize our species' long-standing craving for honey. Several rock shelters in Teruel province in eastern Spain depict outraged bees that seem to issue forth from the rock itself to attack the greedy honey hunters. At Barranc Fondo in eastern Spain, a petroglyph painted about the same time as the one in La Araña shows at least five stick-figure honey hunters climbing a ladder leaning against a large tree. The spirits of large grazing animals and other creatures hover around the top of the tree, lending an air of magic and otherworldliness to the scene. But then in a note of earthy realism, we see one hapless hunter tumbling backward off the ladder, arms flailing in the air. Executed in black, which was derived from charcoal, and red, from ocher, the painting also depicts at least eleven other hunters gathered at the foot of the ladder, waiting for the stitched-hide honey buckets to descend from on high. This scene of hunters anticipating the arrival of their hard-earned plunder is virtually identical to what we will witness in Malaysia when we accompany modern-day honey hunters on an actual expedition.

Graphic depictions of honey hunting can also be found at numerous sites throughout Africa. In fact, Africa has more petroglyphs and associated honey hunter sites than any other continent. Paintings showing bees, hunters, and the ladders, torches, and honey containers necessary to carry out the pillage have been found on rock walls from Algeria, Libya, and Morocco in northern Africa to Namibia, Botswana, and Zimbabwe in the south. Much of this art communicates a sense of the respect, even awe, our ancestors must have felt for the bees they robbed. Their ingenious construction, the honeycomb, is often depicted in extraordinarily accurate detail.

At several sites in Zimbabwe, images of bees were painted in scenes that included antelope, giraffes, and lions, animals whose bones have never been found in refuse heaps in the rock shelters, leading researchers to conclude that they were not eaten and therefore were sacred. Because the bees were shown in close proximity to the sacred animals, it is believed that they too must have possessed some religious significance for the people living in the region.

One of the most important rock art sites in the world is found in the Drakensberg Mountains of South Africa, where something like forty thousand paintings were executed over the course of three thousand years in more than five hundred caves. The paintings were created by the San peoples, hunter-gatherers who settled in the region fifty thousand years ago. The San peoples of today are the direct descendants of these ancient hunter-gatherers. Driven out of the Drakensberg Mountains by Dutch settlers in the nineteenth century, they now inhabit the Kalahari Desert, a vast, inhospitable area that stretches across Namibia, Zimbabwe, and Botswana. Though the modern San are now goat and sheep herders, they have not forgotten the hunting-gathering ways of their ancestors and often supplement their diet with wild game and birds shot with poisoned arrows, and roots and edible bulbs that they dig out of the earth. They also have a taste for honey, and to satisfy it they engage in honey hunts, advancing deep into the bush to locate the nests, smoke out the bees, and devour the fresh honey right on the spot.

An examination of the rock art in the Drakensberg caves reveals a remarkable continuity, linking the rituals immortalized by the San of prehistory to those practiced by the San of today. Many of the prehistoric paintings portray human dancers frolicking below groups of flying bees and hovering bee nests. Similar ceremonial dances still play an important role in San spiritual life. One dance anthropologists have witnessed is performed only when honey bees swarm and could almost be a rock painting brought magically to life. The modern San believe that through this dancing they will be able to harness the force of the bees, a motive that may have been shared by their prehistoric forebears. During the nightlong dances that take place after a successful honey hunt, many of the participants fall into trancelike states, believing themselves transformed into bees and announcing that transformation in a loud chorus of buzzing. We can't be sure, but it's not unreasonable to conjecture that their ancestors fell into similar states after their long-ago bee dances.

A number of the prehistoric San paintings in the Drakensberg caves feature what researchers, for lack of a better term, call formlings, abstract shapes thought to represent sacred beliefs about potency and power. Many of these paintings show larger-than-life bee nests in close association with the formlings, a juxtaposition that may have been thought to ensure bountiful honey hunts.

Paintings depicting honey hunts are also found in caves in India. The rendering of the bees in these paintings is generally less realistic than that in Africa, with the honey makers typically shown only as flying dots. The hunters, however, are considerably more detailed than the African stick-figure hunters. A painting from the Imlikhoh rock shelter in western India shows a right-handed honey hunter wielding a very long, pronged wooden stick identical to those used in the region today to pry honeycombs from inaccessible cliff faces. Streams of bees are flying out of the nest to make short work of the marauder. But the intrepid hunter is not entirely defenseless, for he is brandishing a flaming torch, apparently used as a smoker to pacify the bees. Stupefying bees with smoke is still a common practice among modern-day beekeepers.

The largest prehistoric art site in India is Bhimbetka—sometimes called the Sistine Chapel of rock painting—located about forty-five minutes south of the central Indian city of Bhopal. There are more

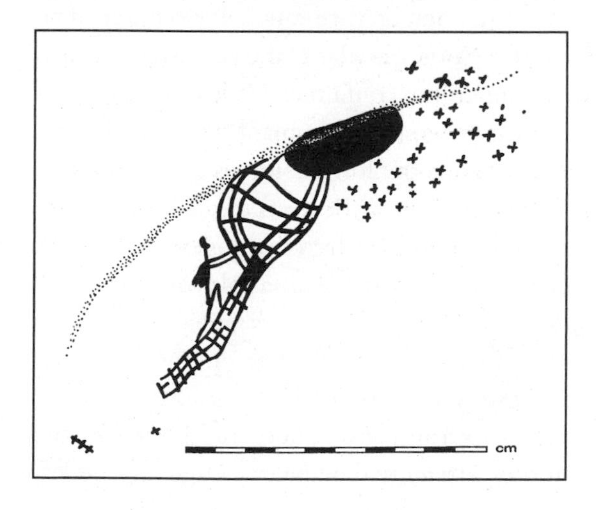

A solitary African honey hunter from the San culture approaches a wild bee nest amidst flying guard bees. Eland Cave, Drakensberg, KwaZulu Natal, South Africa.

than six hundred caves riddling the sandstone hillocks of the region, which is deeply forested with evergreens. The still-vibrant paintings on the cave walls span many millennia, from the Paleolithic through the Mesolithic and Neolithic periods. Most of them are red and white, but some have tints of yellow, orange, emerald green, purple, and crimson. The artists re-created an incredible variety of scenes—hunting parties, wild animal fights, dancing festivals, men riding horses and elephants, and simple household chores such as cooking and tool making, as well as the honey gathering that took place in the surrounding woods. The paintings are detailed enough to give us a remarkably good idea of how the inhabitants of the caves must have lived.

Of course, no one knows for sure why these paintings were made; theories range from simple cave decoration to devotional acts honoring or placating supernatural beings. One theory holds that painting hunting expeditions on stone was part of a cult practice to ensure successful expeditions in the real world. Before leaving on their mission, the hunters may have participated in a ritual during which a successful hunt was drawn in the hope that life would mimic art.

While most honey-hunt petroglyphs found outside of India feature honey bees, which build their nests in dark cavities, those in India record the plundering of rock bee nests. Rock bees generally build one large comb out in the open on the branch of a tall tree. The paintings clearly re-create these nests as well as the prehistoric accessories needed to loot them, including containers made of twigs, dried gourds, and animal hides, ropes woven from plant fibers, and bamboo ladders used to gain access to the lofty colonies. Some hunters are shown waving smoking sticks below the nests to pacify the bees, while others manipulate bamboo poles to free the honeycombs from the surface of the branches, a technique still employed by honey hunters in modern Nepal.

When we look at ancient petroglyphs of bees and honey hunts, we can't help feeling that they are more than mere factual chronicles.

Some of the depictions seem to have a true sacramental quality, conveying the sense of a special relationship between the human hunters and the beleaguered bees they preyed upon. But the precise meaning of what the artists were attempting to communicate has been lost to us. All we can do is conjecture, speculate, and use our imaginations to try to fill in the blanks.

Or we can visit certain contemporary honey hunters such as those in Malaysia and Nepal. Not for them the boxed hives, the heated knives that slice easily through beeswax, or any other of our modern-day contrivances. The rites and rituals they enact during their honey-hunting forays seem to have changed very little since those recorded in the caves. Perhaps by witnessing one of these honey hunts, we will be able to shed some light on the ancient mysteries.

Chapter 2

Searching for Gold: Ancient Rituals and Modern-Day Honey Hunters

His feet are shod with gauze,
His helmet is of gold.
His breast, a single onyx
With chrysoprase inlaid...

—Emily Dickinson,
"The Bee"

We were sitting on a hillside above Pedu Lake in northern Peninsular Malaysia, listening to the mysterious, sometimes unnerving sounds of the rainforest and breathing in its damp, earthy smells. It was a hot, humid night in late February near the end of the rainy season, when the forest comes alive and entire tree crowns burst into riotous pink, blue, yellow, and green. Our group was an unusual mix of cultures, vocations, and interests. It included my longtime friend Paul Mirocha, an accomplished artist who frequently accompanies me on scientific expeditions; his sister, Julie, an independent

filmmaker; several entomologists and evolutionary biologists; and a small party of ecotourists, some of whom had traveled halfway around the world to experience the events about to unfold. We had come together in the old-growth rainforest, with its skyscraper trees and leafy canopy, to witness a traditional Malaysian honey hunt.

We knew that similar honey hunts would be taking place that month in other areas of peninsular Malaysia as well as in Sarawak, Borneo, Brunei, and Indonesia, places that are geographically far-flung but united by the common linguistic and religious traditions of the ethnic Malay culture. They are also united by their common devotion to the fierce giant Asian honey bee *(Apis dorsata)*, maker of a honey valued as much for its reputed healing properties as for its taste. Precious and in short supply, tualang honey (known by the name of the tree in which the bees make their nests) commands a much higher price than honey made by the Asian honey bee *(Apis cerana)* or even by the imported European honey bee *(Apis mellifera)*. The men we would be watching that night risk their lives in pursuit of it.

Distant stars provided the only illumination as we waited on the hillside for the hunters to begin their ascent into the tualang tree, for there were no nearby cities to generate light pollution and no moon to dilute the darkness. An eclectic orchestra of insects and larger animals serenaded us. I was able to pick out the songs of stridulating male crickets, booming toads, trilling tree frogs, and katydids, but there were many other sounds within this rich and unrelenting cacophony that I could not identify. Not far from the spot where we were sitting stood the immense tualang tree where the hunt would take place. Draped with the nests of the giant bees, it soared 240 feet into the sky.

I shifted my weight and settled my haunches into the damp forest soil, trying to get comfortable in anticipation of the long hours ahead. Our guide, Makhdzir Bin Mardan, a professor of zoology at the Universiti Putra Malaysia in Kuala Lumpur, sat beside us, as still and patient as I was eager. Even though this was not the first time I had been present at a Malaysian honey hunt, I could barely contain my excitement. The first honey hunt I had witnessed had been seven years earlier, when I had come to Pedu Lake for a conference on rain-forest honey bees, and it had thrilled me so much that I had repeated the experience many times since.

On the morning of that first hunt, I'd hiked up the steep hill behind the lake to survey the site where the hunters would later be gathering for the night's events, and suddenly caught sight of a towering tualang tree. I knew that the tualang, with its fluted buttresses and smooth, white bark, is the tallest tree in all of Asia. But nothing had prepared me for the actual spectacle of the great tree, which rose 120 feet into the air before the first branches appeared, festooned with over one hundred bee colonies like living Christmas ornaments. A member of the species *Koompassia excelsa,* in the legume family, the tualang is actually a bean tree and would certainly be worthy of co-starring in the tale of Jack and the Beanstalk. Only the enormous redwoods of northern California grow taller.

The giant honey bees that nest in the tualang trees are not permanent residents of the Pedu Lake region. Where they go during their long months away, neither the honey hunters nor modern bee researchers know. They know only that *Apis dorsata* is a migratory

Giant single parabolic comb nests of the largest Asian honey bee (Apis dorsata) *festoon a limb of the tualang tree at Pedu Lake, Malaysia.*

species, always on the lookout for new sources of nectar as forest trees come into flower, then moving on when the bloom begins to wane. Migrating swarms are often seen flying high overhead to their mysterious destinations, moving across the skies like ominous black clouds. But every November, they return to Pedu Lake to make and stockpile honey and to rear a new generation of giant bees.

These bees are truly gorgeous creatures. Their starkly contrasting colors never cease to amaze me. Bold bands of orange-brown, black, and white alternate at right angles to the long axis of their bodies. Their heads and legs are covered with black fuzz. Their wings are also black, with spectacular highlights of blue, purple, brown, and metallic gold, a shimmering palette that could have been created by an impressionist painter. When folded in repose over the backs of the bees, the paired black wings look like the capes of miniature Draculas.

Across the face of their giant honeycombs, six feet long and shaped like half-moons, the bees form living, undulating blankets several layers deep. Beautiful as it is, this layering has a very practical purpose. Unlike the honey bees that we are familiar with in the United States, the giant bees of Asia don't live in tree cavities or other sheltering niches. Their massive but delicate wax combs are exposed to strong winds, intense monsoon downpours, and attacks from birds known as honey buzzards. The dense, interlocking layers of bees covering the surface of the vulnerable nest serve to protect it, keeping the innermost bees, their larvae, stored pollen, and honey safe and dry.

It was during my first trip to Malaysia that I experienced what is perhaps the most startling behavior of the giant honey bees. I was peering through my binoculars at a colony high overhead when I realized that every few minutes the entire surface of the bee blanket covering the comb rippled outward from the center, in much the same way the surface of a pond ripples when a stone has been thrown into the water. It took less than a second for the ripples to reach the outermost edge of the four-foot-wide comb. I was mesmerized by the dazzling display of what appeared to be a perfectly synchronized communal dance. At the time, it was all a mystery to me. But recently researchers have concluded that the bees create the ripple effect to dislodge parasitic flies and wasps that try to lay their eggs on the comb so that their larvae can feed on the bee brood.

The honey-hunting expedition that had drawn us all to the rainforest of Kedah province is part of a tradition that dates back to prehistoric times. Beginning with the reign of their first sultan, Muzaffar Shah I, who ruled from AD 1160 to 1179, the local honey-hunting clans have had to seek royal permission to raid the bee nests. Twenty-seven hereditary monarchs later, Muzaffar Shah's direct descendant, His Royal Highness Al-Sultan Almu' Tasimu Billahi Muhibbuddin Tuanku Alhaj Abdul Halim Muadzam Shah, still has the authority to grant or refuse petitions from the honey hunters in his domain. He has this power because, as sultan, he technically owns everything in the forests.

Following the ritual that had been practiced by his ancestors for eight centuries, Pak Teh, the leader of the hunt we were about to witness, visited the royal palace back in 1965 to plead his case before the sultan. He had just located the bee trees near Pedu Lake and wanted to harvest their honey.

The audience took place in Alor Setar, the capital of Kedah province, at the seventeenth-century Palace Pelamin, a maze of courtyards, high walls, lofty rooms, arched doorways, and steep roofs in the style of Thailand, Kedah's neighbor to the north. In years past, the palace grounds had been the setting of royal weddings and funerals, state visits, and official receptions.

The result of that meeting back in the 1960s was a formal letter issued by the sultan to Pak Teh granting his request to conduct honey hunts. Today, the sultan's letter, suitably framed, hangs in a place of honor inside Pak Teh's home in Jitra, near Alor Setar. It is one of his proudest possessions, he explained through an interpreter as he eagerly showed it to Paul Mirocha and me when we visited him during a previous trip to Malaysia. He went on to describe his audience at the palace nearly forty years before, his eyes still beaming at the recollection. When he had arrived at the gate, wearing his finest sarong with a fresh turban coiled about his head, he was admitted by the guards and led to an elaborate room furnished with silk-cushioned sofas and chairs covered in gold leaf. There the sultan himself was waiting, resplendent in a blue silk tunic, his official robe of office, and the high-peaked royal turban, with its jewel-encrusted ornamentation. The sultan's tunic was covered with large, gleaming medals, and his ceremonial dagger, thrust into his wide belt, hung by his side.

After the formal greetings, Pak Teh made his oral petition to the sultan, who in reply invited him to sit and describe a honey hunt. When Pak Teh finished speaking, the sultan smiled and gave him permission to harvest honey in the protected forest around Pedu Lake. Every year since then, a letter from the sultan has arrived at Pak Teh's home renewing his honey-hunting privileges—just as every year at the end of the hunt, Pak Teh returns to Alor Setar, where the sultan now lives in a new palace, to deliver a small amount of honey as tribute to the ruler. (The old seventeenth-century palace, by the way, has been made into a museum housing artifacts from the reigns of earlier sultans. It is a fascinating place to visit, as I myself can attest.)

The sultan's annual letter confirming Pak Teh's right to harvest the Pedu Lake honey arrived at his home two weeks before the hunt that we had come to witness was to take place. Not long after its arrival, Pak Teh began preparing for the event. Pak Teh is not only the leader of the honey hunt, but also the oldest member and leader of his clan. All of the other hunters are related to him: cousins, brothers-in-law, nephews, and grandsons. He is a compact, wiry man, with a quick, easy smile and a bald head wrapped in a turban fashioned from a towel. When not planning and leading honey hunts, he keeps busy as a rice farmer, rubber planter, carpenter, and Imam of his neighborhood mosque.

Pak Teh and the other members of the clan spent several days checking and repairing their equipment—the climbing ropes, honey containers, leather buckets, bone knives, and liana torches that they would need for a successful harvest. (Liana torches are made from sturdy vines that colonize the rainforest trees. After being pounded with knives and reduced to soft, pliable fibers, the lianas are bound into six-foot-long bundles about four inches thick. Each torch can burn for an entire night.) A week before the hunt was to begin, they packed the equipment into their old blue minivan and set out on their journey. The team was made up of Pak Teh, four other adult men, and Pak Teh's two teenage grandsons, Shukor and Nizam, who, in an important rite of passage, were going to be initiated into the honey-hunting rituals.

When they reached the rainforest site an hour later, the hunters set up camp about fifteen minutes downhill from the bee tree, next to a small, clear stream. They hung a large blue plastic tarpaulin from a rope strung between two trees to serve as their roof, then spread a blanket on the ground below the tarp. A few sheets and pillows made up their simple beds. Lashed railings fashioned from cut saplings provided walls for their temporary home. They then unpacked their cooking pots, plates, and utensils and arranged them on the rocks that formed the campfire circle, a semipermanent installation used year after year. Finally, the hunters dammed part of the stream to create a pool for bathing.

The low-growing plants near the camp were spotted with thousands of small yellow dots, remnants of feces the bees released on their evening flights out of the nearby nests. On an earlier trip, I had collected some of this material and analyzed the pollen grains it contained in order to identify which flowers the secretive bees had visited high in the forest canopy. On this trip, I was planning to collect honey and pollen from the combs as well as a new supply of bee feces to analyze back home in Tucson.

Pak Teh was optimistic about the upcoming harvest. Ever since the bees had returned from their annual migration the previous November, he had been making occasional visits to the rainforest, hiking to the bee trees to observe how many colonies had arrived and how the flowering season was progressing. In this way, he was able to predict whether the harvest would be bountiful or not—and happily, this year his predictions were encouraging.

Once they were established at the camp, the seven honey hunters devoted most of their preparations to replacing sections of the old ladder that was attached to the bee tree. The humidity of the rainforest as well as hordes of voracious termites had taken their toll on the structure, making it unsafe. The men used nails to reanchor the ladder to the trunk of the tree and reinforced the crosspieces with sapling wood and rattan, which had been gathered in the forest. After nearly a week of hard work, all the equipment had been readied, the climbing ropes neatly coiled and the large leather and rattan buckets scrubbed and placed conveniently at hand. To facilitate the raising and lowering of the honey buckets, a wooden pulley had been secured to the underside of a massive limb 120 feet high in the tualang tree.

After a long hike from the main road, our group had arrived at the forest camp on the afternoon of the day the hunt was to begin. With several hours of light left, we were able to watch the hunters as they made a final check of their nylon ropes and long, tightly bundled liana torches. As evening approached, Makhdzir Bin Mardan cleared his throat and prepared to tell us about the traditions governing the honey harvest so that we would better understand the ceremonies we were about to witness. Professor Mardan had become an expert in the honey hunt through repeated trips to these forests, interviewing the hunters and studying the Hindu, Islamic, and animist traditions that form the basis of their rituals. Born in Malaysia but educated at Guelph University in Canada, he is a slender man with glasses, black hair, and a thin goatee, as much at home in the rainforest with the giant bees as he is on campus and in lecture halls.

Out of the vast store of tales filed away in his memory, Makhdzir chose the ancient fable describing the origins of honey hunting in this part of the world.

"Long ago," he began, "a princess of the royal family had a Hindu handmaiden, a dusky beauty called Hitam Manis or 'Sweet Dark One.' The handmaiden fell hopelessly in love with the sultan's son, a handsome prince who requited her passion. But their love was doomed, for she was a commoner, and marriage of a commoner to a prince of the blood was strictly forbidden. When the sultan learned of the romance, he flew into a rage, and Hitam Manis, along with the other handmaidens, the Dayang, had to flee the palace for their very lives. As the terrified young women escaped into the forest, they were pursued by the sultan's guards, who hurled long metal spears at them. When one of the spears pierced the already broken heart of Hitam Manis, miraculously she did not die. Instead, she and the other handmaidens were transformed into a swarm of bees and disappeared into the night. Thus were born the giant honey bees of the Asian rainforests."

Pausing in his story, the professor suggested that we lie down on the forest floor and look up into the canopy of the tualang tree in anticipation of what would happen next in the story of Hitam Manis.

"Years later, the still grieving prince—now engaged to a proper princess—noticed a large honeycomb high in the branches of a tualang tree in the forest. When he climbed the tree to investigate, he discovered a large cache of golden honey. He called down for his servants to send up a metal knife and bucket so he could harvest the treasure. The servants dutifully sent the knife and bucket up to the prince, but when they lowered the now heavy pail a few minutes later, to their shock and horror, they found the prince's dismembered body inside.

"From the treetops, a ghoulish voice cried out that he had committed a sacrilege by cutting the honeycomb with a sharp metal knife. Unwittingly, the prince had insulted poor Hitam Manis, reminding her of the cold metal spear that had pierced her heart and so changed her life.

"But the Sweet Dark One took pity on the prince she had once loved, and released a golden shower that restored him to life and limb."

Professor Mardan went on to explain that "golden showers"—

which leave thousands of golden spots on the forest foliage like those found near the honey hunters' base camp—are actually mass defecations made by *Apis dorsata* during their flights after sunset, when they rid themselves of feces and unwanted heat. During the Vietnam War, American GIs had mistaken these golden showers for the dreaded yellow rain, a form of biological warfare that poisoned thousands of Vietnamese villagers. Thanks to the work of Paul and Julie's father, biochemist Chet Mirocha, as well as other scientists, we now know that golden showers are a natural biological phenomenon—and an ecologically valuable contribution to the environment. In fact, Malay farmers consider themselves lucky to have fields or rice paddies near a bee tree. Not only is it a good omen to have the bees living nearby, but the farmers know that the daily cleansing flights deposit great amounts of nitrogen- and pollen-rich feces on their crops.

"To this day," Makhdzir concluded, "in deference to the dying anguish of the handmaiden known as Hitam Manis, honey hunters never use tools made of metal—only those of wood, cowhide, and bone."

And so it is that Pak Teh and the other honey hunters of his clan use leather buckets and knives made from the shoulder bone of a cow.

In the early evening, not long after Professor Mardan had recounted the Hitam Manis story, Pak Teh assembled his hunters in a semicircle at the edge of the camp. As our group watched from a respectful distance, he gave instructions to the four climbers, who included his grandsons Nizam and Shukor, and then to the two older men who would serve as the ground crew, sending buckets up to receive the giant beeswax combs and lowering them to be emptied when they were full.

Next Pak Teh cupped his bronzed, wrinkled hands around a smoky oil lamp and uttered an ancient prayer, the same one he had been reciting in this very spot for nearly forty years. We couldn't understand the Bahasa dialect spoken in this border region, but that didn't matter, since the words were not meant for our ears. In fact, it was forbidden for us to hear what was being said. We were not members of the honey-hunting clan, nor had we earned the right to participate in their ancient rituals. We were outsiders.

Makhdzir explained to us that Pak Teh was asking the tree, the forest spirits, and the giant bees themselves for permission to make the night's climb and for a successful harvest. His words were drawn from Islamic and Hindu prayers as well as animist beliefs dating back to the time when only the indigenous Orang Asli tribes lived in these forests. We could imagine ancient Orang Asli honey hunters uttering similar prayers thousands of years ago, long before the coming of Pak Teh's people, not to mention the nosy bee scientists and ecotourists from half a world away.

After Pak Teh had finished the ritual prayers, our group was led from the base camp to the bee tree clearing on the hillside. Because night had already fallen, we didn't have to worry about the bees, who couldn't see us in the darkness. People who visit bee trees during the day, however, have to walk slowly and use the branches of tall shrubs to hide their movements. Anyone who doesn't take such precautions is likely to suffer countless stings delivered by the nest's fierce guardians. In fact, it is not uncommon for people to be hospitalized as a result of these massive, sometimes fatal attacks. Still, despite the risks, the Malay honey hunters always refer to the giant honey bees with great tenderness. They call the bees Hitam Manis, a lover worthy of a royal prince. During the hunt, they humbly refer to themselves as Dayang, the palace handmaidens of legend. The few times I observed Pak Teh being stung, he always smiled as he gently brushed off the offending bees. As far as he was concerned, he might have been getting bee kisses from his fine friends, instead of painful stings from the world's largest and fiercest honey bee species.

We were lucky that there was no moon on the night of our hunt. On nights with a moon, the hunters have to wait until after it has set to avoid detection by the resting but ever vigilant bees. Now, as we watched from our hillside vantage point, Pak Teh and his fellow climbers quickly started their ascent with coiled ropes slung over their shoulders. They climbed the ladder barefoot, wearing loose clothing with no Western-style veils or protective beekeeping equipment. Shukor trailed the long liana torches from a cord attached to his waist. The hunters climbed in silence and without flashlights until they reached a height of more than one hundred feet. (Flashlights might alarm the bees and set off a defensive attack.) The first level branches were four or five feet in diameter. Beneath them hung the massive combs, each six feet long, four feet wide, and fat with honey. Both sides of the combs were covered with striped bees in interlocking layers five or six deep.

It was a familiar climb for Pak Teh. He had been harvesting honey from the Pedu Lake bee trees every year since 1965. Now, with his eyesight failing due to cataracts, he was eagerly training his grandsons in the important traditions of the honey harvest. Though this was their first hunt, Nizam and Shukor were scrambling up into the branches of the tree with as much agility and fearlessness as their much more experienced elders. The rest of us sat below on the damp hillside, waiting and listening for any sounds from above.

During my first visit seven years earlier, I had hammed it up, bravely climbing the ladder a modest fifteen or twenty feet for photo opportunities to impress friends and family back in Tucson. Climbing the rest of the way up to the bee nests was unimaginable. The honey hunters, however, are much braver. During the ascent, they scamper up the sturdy but frail-looking ladder with remarkable speed, seeming entirely at ease with this death-defying feat.

When they reached their precarious treetop destinations, it was time for the harvest to begin. Pak Teh's voice rang out as he announced that he was poised on a branch immediately above a bee nest, ready to cut the honeycomb. But before the comb could be cut and folded into the leather bucket, the bees would have to be evicted from their home—a process that involved the clever use of both fire and song. Fifty feet from where our group was sitting, Pak Teh's brother-in-law waited at the base of the tree. Now, in a loud, clear voice, he began to chant, calling to the tree, the stars, the spirits of the forest, and the bees themselves in their lofty nests. His voice rang strong and true through the moist, hot air of the tropical night. Even the noisy insects seemed to grow quiet when he sang his song of Hitam Manis.

> *"Hitam Manis Ooooi!"*
> *(Sweet Dark One, Ooooi!)*
> *"Turunlah dengan chahaya bintang"*
> *(Come down with the falling stars)*
> *"Turun dengan lemah lembutnya"*
> *(Come down gracefully)*

A herringbone ladder snakes its way 120 feet up into the boughs of the tualang tree climbed by Pak Teh and his honey-hunter clan. Pedu Lake, Malaysia.

When Shukor passed the burning liana torch to his grandfather, we saw its glowing tip arc through the still night air. Soon a cascade of orange embers rained down like a meteor shower from the branches overhead. No Fourth of July fireworks display has ever been so memorable for me. It is a pyrotechnic spectacle that has kept me returning to the bee trees of Pedu Lake year after year.

As the first orange sparks floated lazily to the ground, I heard an ominous roar from above. The inhabitants of Pak Teh's colony were flying in our direction—tens of thousands of incensed bees following the rain of fire earthward. On my first honey hunt years before, I had instinctively ducked and huddled in the darkness, shaken by the prospect of imminent attack and convinced that the bees would easily find me. But now I knew there was nothing to fear. Long ago in these ancient forests, honey hunters learned how to manage the bees using the sparks from burning liana torches. The bees obediently followed the drifting trails of glowing sparks to the ground. The roar of the oncoming bee locomotive ended as abruptly as it had started as the bees settled harmlessly on the vegetation below. The bees would be unable to find their way home or attack the honey hunters until the morning light. But by then the hunters would be safely out of the tree.

Pak Teh's brother-in-law repeated the chant to Hitam Manis whenever he saw sparks spilling from the torches above, the signal that the climbers were driving the bees from yet another nest. The distinctive rhythm of the chant mesmerized us, filling the tree clearing as it cajoled the bees to leave their colonies high in the tualang branches.

As the raid proceeded, the giant waxen combs were cut from the tree branches, folded into halves or quarters, and sent down in the leather buckets to the two men waiting on the forest floor below. After the buckets had been emptied and sent back up to be refilled, the combs were squeezed through funnels covered with cheesecloth into two immense blue plastic containers, each holding perhaps a hundred liters of honey. Finally, at about four a.m., after seven hours in their treetop workplace, the climbers made their way down to solid ground, having collected honey from a dozen or more colonies. (Each night for a week they would climb again into the tualang tree until over a thousand pounds of honey had been harvested from perhaps eighty colonies.) Visibly exhausted from the long night's exertions, the hunters nonetheless had an air of exhilaration about them, the look of men who know that they have successfully completed a particularly difficult job.

But Pak Teh still had one more important task to perform. Standing at the foot of the towering tree as the rising sun began to color the sky a brilliant crimson, he prepared to carry out yet another time-honored ritual. As his weary men looked on, he carefully selected a large honeycomb from those in the honey bucket. Lifting it, he uttered something we couldn't hear, then hurled the comb deep into the forest behind the tree. Makhdzir explained that this, the first honey taken, was an offering to the "unseen owner" of the forest and its trees, given in thanks for yet another safe and bountiful harvest.

Our group now joined Pak Teh and the others under an immense boulder overhang where, tired and hungry, we all huddled around the sticky harvest to partake in a ritual meal of honey and brood. It was a joyous celebration of the life-giving nectar and pollen of the forest. Some of us vied for the tastiest morsels, letting the honey drizzle down our throats, then spitting out the indigestible chunks of beeswax. Pak Teh handed me a piece of comb heavy with white bee grubs peeking out of their cells and urged, "*Makan, makan*" ("eat, eat"). Not wanting to offend him, I tasted the plump bee larvae. My UC Davis graduate studies had taught me that honey bee larvae make a wonderful quiche, and I now discovered they weren't bad raw.

Meanwhile, under the boulder overhang, the men from Jitra continued to laugh and joke as they filtered honey into the big blue plastic containers. In a few days they would break camp and drive out of the Pedu Lake forest. Back home, they would have another ritual feast with their families and friends. Once again they would give thanks to the forest and its majestic bees. And then they would divide the honey among themselves, keeping some for their own use, then bottling and selling the rest in neighboring markets or to the local honey cooperative.

Honey hunting is in the blood of Pak Teh and the others of his clan. What seems to us like the most dangerous of pursuits is an event that they look forward to and cherish each year. While the supplemental income from selling the tualang honey is important, what matters most is the special relationship they have with the rainforest bees.

After the hunt was over and the men had returned to Jitra, Paul, Julie, and I visited Pak Teh's home to photograph the clan and sample more of the marvelous tualang honey. Built over one hundred years ago by Pak Teh's father with termite- and rot-resistant rainforest timber, the house is the oldest building in the oldest part of town. It is held together with wooden pegs and rises eight feet above the ground on sturdy wooden stilts, insurance against the monsoon floods that frequently inundate the region. The wooden walls and roof beams have been elaborately hand-carved with intricate geometric and floral patterns. The weather-beaten old house sits proudly in a big garden planted with lush fruit-bearing trees.

Before leaving Jitra, we asked Pak Teh if honey hunting would continue into the future. After all, during this hunt, his grandsons had been literally handed the torch. In response to the question, Pak Teh just smiled and uttered the soft, pleasing laugh we had heard so often during the past few days.

"As long as there are tualang trees in the forests," he said, "and giant bees to build nests in them, there will be hunters like us to harvest their honey."

Honey Hunting in the Shadow of the Himalayas

Modern-day honey hunts that carry on ancient rituals and traditions are not confined to the rainforests of Malaysia. Honey is still harvested by intrepid hunters in many far-flung parts of the world, from the valleys of Nepal to the homeland of the Australian Aborigines.

In Nepal, our friends the giant honey bees build their nests not in towering trees but on sheer cliffsides in order to discourage predation by honey-hungry humans and other animals. Situated at the base of the Himalayas, Nepal provides an abundance of these seemingly impregnable nesting sites. But even so, there is no guarantee that the bees' refuge won't be breached by human ingenuity, daring, and greed, as has been happening for thousands of years.

Many men of the Gurung tribe, who live in the foothills of the Himalayas in west-central Nepal, still practice the art of the traditional honey hunt, risking life and limb as they dangle on rope ladders high above the ground to loot the giant combs attached to the cliffside.

Although no historical records exist, Gurung honey hunts have almost certainly been taking place for millennia, with generation after generation of hunters passing their ancient skills on to their sons, teaching them all they need to know, from how to perform the ritual offerings that will protect them from danger and death to the correct way to light and maintain the smoky fires that will pacify the fierce bees.

The hunting expeditions, generally made up of five or six men, can last a week or more. It may require two or three days to complete the journey from the hunters' isolated villages to the remote bee-nesting cliffs where the treasure awaits. Barefoot and with little in the way of warm clothing to keep the cold mountain air at bay, the men negotiate a network of narrow paths worn deep by centuries of use, making their way through a landscape of steep, terraced hillsides planted with rice, millet, and other crops, with the snow-capped Himalayan peaks soaring on the horizon. One unlucky hunter is designated to carry the 160-foot-long rope ladder, which can weigh over seventy-five pounds.

The hunters' faith is deeply embedded in both Buddhism and Hinduism, the principal religions of Nepal, though the influence of ancient animist beliefs is still powerful, allowing them to maintain a close relationship with their local deities. Soon after they set out on their journey, the lead honey hunter generally leaves the group and selects two small bundles of branches from a medicinal plant. He places one bundle under a stone in the middle of the path to prevent any village witches from following them. The other bundle he uses to tap each man on the shoulder while he recites ancient mantras of protection in a hushed voice.

As soon as they reach the bee-nesting cliff, the leader prepares an offering, shaping the figure of a local mountain god out of thick millet paste and placing the effigy on a stone slab along with other offerings, including rice, millet, popcorn, strands of sheep's wool, and juniper twigs. As he murmurs incantations to the deity, he sprinkles corn over the slab and sets the juniper twigs afire. Finally, he sacrifices a live chicken in the hope that the god will spare their lives.

After the ritual has been completed, the sacrificed chicken is cooked in a spicy curry sauce and eaten. Then the men go to work, cautiously lighting fires near the base of the cliff. When the bees smell smoke from the fires, they think their nest is in danger and gorge on honey in order to have plenty of reserves in case they have to flee. It's a smart move on the part of the marauding villagers, since smoke-dazed, overfed bees are less likely to sting.

Once the fires have been lighted, the men climb around to the top of the cliff above the bee nests. They anchor the rope ladder to a tree and push it off the edge, high above the smoking flames. Slowly the leader of the hunt begins his descent.

The bee nests, attached to the face of the cliff, are usually as wide as a man is tall and several feet deep. The lowermost crescent of the parabolic comb is home to thousands of immature bees. Suspended precariously from his ladder and maneuvering two long bamboo poles like giant chopsticks, the lead hunter cuts away the lower portion of the comb with great precision and allows it to crash to the earth below. The wax will later be rendered and the nutritious larvae eaten by the team members. The leader then uses the poles to detach the uppermost part of the comb, where the honey is stored, and eases it into the honey bucket. But first, the dark mass of resident bees has to be chased off. To accomplish this, he passes flaming bundles of leaves beneath the nest. Instantly, a swarm of loudly buzzing bees rises from the comb and flies away, allowing the climber to collect the honey in relative safety from his perch high above the rocky gorge. Although the nests are destroyed during the harvest, the bees generally return to the remnants and build new nests in as little as a month.

Each nest yields from 135 to 160 pounds of honey and beeswax, both of which are valuable commodities in the marketplace. The Nepalese use honey not only as a food but also as a universal remedy, believed to be good for whatever ails you. It is sold to villagers or exchanged for other products, such as milk, yogurt, grain, or perhaps a chicken. The wax too has its uses, for many artisans in Kathmandu still practice the ancient art of lost-wax casting when making religious figurines and are eager to purchase the hunters' stock.

While Nepalese honey hunts are still a viable source of both honey and income (not to mention adventure) for the impoverished villagers, I can't help wondering how many more there will be. As in Malaysia and so many other parts of the world, the forests of Nepal are shrinking rapidly as more and more wood is cut for fuel and building materials. And when the trees disappear, the torrential monsoon rains, with nothing to block their way, gush down the hillsides, washing away both meadows filled with wildflowers and terraced paddies, important sources of nectar and pollen for the bees. As their raw materials dwindle, compromising their ability to produce honey, increasing numbers of bees are abandoning their cliffside dwellings and relocating in parts unknown. How long the others will remain is a question no one can answer.

The Sugarbag Quest: Honey Hunting in Northern Australia

For thousands of years, the Aborigines of Australia have survived in a landscape whose harsh conditions would have sapped the will and broken the spirit of many other peoples. In scrubby, sun-baked deserts, they learned how to harvest what they needed from a seemingly ungiving natural world—and one of the things they needed was honey. Like all the other human inhabitants of our planet, Aborigines have a serious sweet tooth, and to satisfy it, they have long plundered honey from the nests of their native bees, particularly stingless bees of the genera *Trigona* and *Austroplebeia*. Unlike Pak Teh's clan and their Nepalese counterparts, the Aborigines don't have to worry about attacks from outraged defenders of the nests they are violating, for bees without stingers are much less formidable than their well-armed relatives.

Aborigines use the English word *sugarbag* to refer to both the hard-working stingless bees and the honey and resins they loot from the bees' nests. The honey plays an important role in their diet, improving the flavor of many foods and providing a sweet drink when mixed with water. As in many cultures around the world, it also sweetens a number of Aboriginal social rituals. If a young man seeks a bride, he

*A group of Australian Aborigines gather around the "sugarbag," a nest of stingless bees (*Trigona *spp.), enjoying the sweet prize.*

gives honey to her parents to encourage a favorable response to his proposal. When one tribal group calls on another, proper etiquette requires that they bring along a gift of honey. Honey is also used to treat a number of ailments, including burns, lung complaints, sore eyes, vomiting, and diarrhea.

Nor does this exhaust the uses the Aborigines find for the products of the bee. Resins plundered from bee nests in the wild are still used as glue in tool making and as a sealant for bark buckets. And strips of resin are pressed onto rock surfaces to form human, animal, and spirit figures in the creation of petroglyphs, an art the Aborigines still practice.

The Aborigines' affinity for sugarbag is rooted in the shadows of prehistory, as we know from ancient cave paintings found in northern Australia, which frequently portray honey bee nests in cross section, detailing both the entrance and the honeycombs. Often human hunters are shown poking sticks into the entrances as the bees, indicated by angry red dots, swarm out in a retaliatory rage. Some of these prehistoric sites are now associated with "honey bee dreaming."

The ancient Aborigine tradition of dreamtime isn't about dreaming at all. It is actually a rich, complex collection of oral myths and legends that explain the creation of the earth and the origins of the plants, animals, and people that inhabit it. Recounted by tribal elders around blazing campfires, these stories help Aborigines reaffirm their connection to the past as well as to the natural world in which they live. Honey bee dreaming is a repository of the traditions that govern the ways humans think of and interact with honey bees.

According to one honey bee dream, ancient honey hunters devised an ingenious way to locate bee nests in the wild. The hunter would deftly attach a feather or the filament of a spiderweb to the hind leg of a forager bee, using the gummy resin secreted by a particular herb. The bee, feeling the weight and mistaking it for a cargo of nectar or pollen, would head home, thinking her workday was over. Flying low to the ground due to the extra burden, she made it easy for the keen-eyed hunter to follow her back to the honey treasury.

Whether or not this tactic was ever actually practiced, it underscores the fact that stingless bee nests can be very hard to find in the wilderness of the northern Australia outback. Unlike our familiar honey bees, stingless bees are shy, unwarlike, and unarmed, but their secretive ways help them compensate for their lack of defenses. When they sense danger, usually in the form of a greedy human or animal predator, they pull back into the nest, where all activity ceases until the threat has passed. The elusive bees, however, are no match for the master Aboriginal honey hunters of today. Drawing on the accumulated knowledge of their honey-hunting forebears, they are expert at locating bee nests amid the tangle of forest trees. They know which trees are preferred by the bees and where those trees are likely to grow. When they find a tree that looks promising, they press their ear against the base of the trunk and listen carefully. If they detect a faint humming sound from above, they know they have struck liquid gold. Another giveaway is the presence of certain small black lizards that live near the nests and dine on the returning bees, snatching them out of the air with their sticky tongues.

When a bee tree has been identified, the hunters immediately go to work, wielding stone axes to cut notches into the trunk to serve as footholds during their ascent. Up they go until they reach the first branches, high above the ground. By tapping the branches with the back of the axe, they are able to determine exactly where the nest cavities are. Once they have located a nest, they use the axe to widen the entrance so they can reach in and scoop out the honey (remember, stingless bees can't sting). After helping themselves, the hunters use mud to repair the breach they've made so that they'll be able to return for a refill in the coming weeks or months.

But how long these honey hunters or any others will be interested in returning remains to be seen. While some Aborigines still practice the ancient art of the honey hunt, the encroaching influence of non-Aboriginal Australia is taking its toll. Younger generations are losing touch with their roots in dreamtime, seduced by the glamour and technologies of a global culture that is focused on the future to the detriment of the past. Urbanization, with its pitfalls of drugs, alcohol, and consumerism, has undermined the Aboriginal understanding of their land, the special connection that was embodied and preserved in the legends and lessons of dreamtime.

Now that we have come to know the ways of traditional honey hunters, it's time to pay a visit to those who have taken an easier and less risky route—keeping bees in man-made hives they can plunder at will.

Chapter 3

Staying in Touch: The Beekeeper's Craft

The little bee returns with evening's gloom,
To join her comrades in the braided hive,
Where, housed beside their mighty
honey-comb,
They dream their polity shall long survive.

—Charles Tennyson Turner,
"A Summer Night in the Bee Hive"

While honey hunting provides adventure and a connection to age-old traditions, beekeeping is a lot easier and more practical, and it has a long history as well. The keeping of bees is but one aspect of the indefatigable human urge to subjugate nature. We've tamed wild dogs and cats, cattle, and horses, turning them into obedient pets, docile milk producers, and sturdy mounts. We've rerouted mighty rivers, plowed under ancient forests, drained swamps and reinvented them as gated communities and golf courses. So it's hardly surprising that we domesticated bees, bending them to our will, channeling their natural honey-making instincts to our own purposes,

whether it be to satisfy our urge for sticky buns and honey wine or to create healing unguents and preservatives.

Beekeepers are not your average citizens. It's their passion for bees that sets them apart. Most people in Western countries, especially the United States, have pronounced entomophobia, or fear of insects. A tiny creature buzzing around the head seems to send most normal folks packing, or at least reaching for a can of bug spray.

Not so with beekeepers, for beekeepers have an intuitive understanding of their bees, can sense their moods, predict their actions, and anticipate their every need. They have even developed unique ways of communicating with their fuzzy, honey-producing charges—a private language of words, sounds, and song. To a degree, the relationship is about commerce and frank exploitation—but it's also about so much more. There is a magical, almost sacramental quality that binds beekeepers and bees, a (dare I say it?) deep and abiding love.

But when did this strange passion called beekeeping begin? About seven thousand years ago, the trend toward a sedentary life changed the relationship between humans and honey bees. In addition to plundering their nests in the wild, we began raising them on our farms and in the backyards of houses in our villages.

The first beekeepers may have simply pried loose the inhabited sections of bee trees and carried them back to their settlements to protect them from the ravages of other honey-loving predators. Centuries later, the Aryan Indians and the Egyptians refined the process.

Beekeeping in India can be traced back to the Vedic period, beginning four thousand years ago, when the Aryan invaders of the subcontinent amassed considerable knowledge of bees and their honey-making ways, much of it remarkably accurate (though, like most early beekeeping peoples, they believed each colony was ruled by a king instead of a queen). Honey was used not only as a cooking ingredient in Vedic India, but also as a medicine with wide-ranging therapeutic applications. It also played a significant role in Vedic myths and religious rituals. During marriage ceremonies, for example, the bride's body was anointed with honey to ensure fertility. And after the ceremony was over, honey was served to the guests because it was believed that a substance so pure would ward off any evil spirits that might try to crash the festivities. With honey on the menu and evil spirits kept at bay, the newlyweds could expect to have a happy, fruitful, and prosperous married life.

Vedic beekeepers housed their honey makers in hives constructed of twigs and grasses covered in dried mud, as well as in clay pots. The hives were typically kept in wall niches or hung from the ceiling in farmyard outbuildings. As is the case in many parts of the world, both historically and in the present day, Indian beekeepers used smoke to pacify their bees while they helped themselves to their honey. They were, however, wise enough to leave sufficient honey in the hive to ensure that the bees remained well fed and productive. Overpopulated colonies were often divided to create new ones—still a common practice among beekeepers around the world. Empty hives were somctimes hung from trees in the forest and smeared with beeswax and sweet palm sap to encourage swarming bees to settle down and get to work. Once the bees had established themselves in their new home, the beekeeper would bring the now populated hive back to the farmyard so that the honey could be conveniently harvested.

Beekeeping remained widespread in India until about AD 200, when the cultivation of sugarcane and the refinement of sugar processing methods led to the displacement of honey as the preferred sweetener. Another reason for the decline of honey may have been the spread of Buddhism and Jainism, two religions that prohibited their adherents from depriving animals of either their lives or their food supplies. Beekeeping and honey harvesting may have been regarded as sinful, and as a result, they fell out of favor.

Lower Egypt, well watered and fertilized by the annual flooding of the Nile, seems to have been the center of organized beekeeping in the ancient world. In fact, bees and honey were so important to the economy of Lower Egypt that the honey bee hieroglyph was chosen as the symbol of the entire region.

An incised Egyptian tomb hieroglyph of a bee, the symbol of royalty. From a door lintel, King Intef.

Ancient Egyptian beekeeper pouring honey into a container. Incised and painted relief on a stone pillar in the tomb of Pabesa, West Bank, Upper Egypt, 664–625 BC.

Temple paintings and reliefs dating from 2400 to 600 BC depict the harvesting, processing, and storing of honey. The lightweight Egyptian hives were ingeniously simple—wicker baskets covered with clay and baked in the sun until hard. This meant that, for a fee, they could be easily transported from one farmer's field to another to facilitate pollination. Large earthen vessels were used for honey storage. As in India, the honey produced by the hives' resident bees had ceremonial as well as culinary uses. Egyptian gods apparently craved honey, and to satisfy the gods' inexhaustible demands for ever larger offerings, priests resorted to keeping their own hives within the temple precincts. The honey from the temple apiaries, believed to possess special healing properties, was used in the manufacture of many medicines and ointments.

Honey wasn't the only reason to raise bees. Always practical, the Egyptians put beeswax to good use in a number of ways, including mummification, shipbuilding, the lost-wax casting of sacred objects, and as a kind of gel to keep their elaborate wigs slicked down and firmly in place. During certain execration (cursing) rituals, beeswax figurines representing enemies of the state and occasionally sexual rivals were thrown into roaring fires, where they instantly melted—a sure way to render the offenders impotent in both the political and sexual arenas.

Egyptian beekeeping practices influenced successful bee management throughout the entire Mediterranean world. We know that the Greeks and Romans were avid beekeepers and consumed honey in great quantities. In fact, the Romans elevated beekeeping to a fine art—so fine that the great Roman poet Virgil wrote about it in lyric verse. In book four of the *Georgics,* his famous treatise on agriculture and beekeeping, Virgil covered just about everything a beekeeper needed to know, from the ideal location of the apiary . . .

> *Let there be clear springs nearby, and pools green with moss, and a little stream sliding through the grass.*

. . . to its proper maintenance . . .

> *You keep them warm too, with clay smoothed by your fingers round*
> *their cracked hives, and a few leaves on top.*

. . . to how to control a swarm of bees on the wing . . .

> *Scatter . . . balm and corn parsley's humble herb and make*
> *A tinkling sound [with cymbals]:*
> *They'll settle themselves on the soporific rest sites:*
> *They'll bury themselves, as they do, in their deepest cradle.*

Virgil also praised the communal values of honey bee society:

> *They alone hold children in common: own the roofs*
> *of their city as one: and pass their life under the might of the law.*
> *They alone know a country, and a settled home,*
> *and in summer, remembering the winter to come,*
> *undergo labour, storing their gains for all.*

Virgil's glimpses into life in the hive advanced the notion of bee society as a model that human society would do well to emulate.

Organized beekeeping, which had thrived and spread throughout the entire Roman Empire, went into a decline with the demise of the Pax Romana. During the Dark Ages in Europe (AD 500 to 1000), and especially after the ruinous invasions of the eastern hordes, apiculture nearly ceased in many parts of the continent. Most people had to make do with hunting honey in the forests. Toward the end of the Dark Ages, Charlemagne, the first Holy Roman Emperor (AD 800 to 814), laid down rules governing the beekeeping that was still practiced—rules having mostly to do with taxation to fill the royal treasury. Beekeepers were obliged to pay the emperor dues in kind: two-thirds of all their honey and one-third of their beeswax. After Charlemagne died, *abeillage,* or "bee dues," remained a feudal right. Every vassal owed the sovereign a generous portion of what the hives produced.

Even before their colonization by Rome during the reign of Emperor Claudius (10 BC to AD 54), the ancient Britons were so famous for their beekeeping skills that the seafaring Phoenicians, who regularly visited the island on trading expeditions, referred to it as the "Isle of Honey," as did local Druid bards. Under Roman rule, the country produced and consumed vast quantities of honey, both as a sweetener and in the form of mead, the honey wine that was their beverage of choice in the days before neighborhood pubs. Centuries later, at the end of the Dark Ages, the Domesday Book, a record of the great survey of England that was completed in 1086 for William the Conqueror, mentions that a goodly number of managed beehives could still be found throughout the kingdom. After milling, fishing, and mining, beekeeping was the most prevalent industry in the land. Unlike their Continental counterparts, English beekeepers apparently prospered throughout the Dark Ages.

The Middle Ages (or medieval period) in Europe began around AD 1000, when law and order had been restored following the barbarian invasions, and lasted until the fourteenth century, when the Renaissance inaugurated a new era of rational thinking. By the late Middle Ages, the feudal system of land tenure, which had become firmly entrenched in the absence of a strong, centralized government, had given way to the growth of national identities and the rise of states such as England, France, Denmark, and Norway.

Throughout this period, apiculture, or the tending of honey bees, was a common practice at monasteries across the Continent. Hives were built into special niches in cloister walls or placed in bee gardens, where they were tended more or less in the wild.

As the Catholic Church became more and more dominant, the need for beekeepers increased. Whenever mass was sung in the growing number of churches and cathedrals, only candles made of pure beeswax produced by "virgin bees" could be used. (Until the fifteenth century, the Church fathers thought that bees were chaste creatures and, like Christ, came into being as the result of virgin births, without stooping to "unseemly sexual acts.") Bees, pure and sinless, were believed to have fled the Garden of Eden when Adam and Eve fell from grace, and therefore they were creatures blessed by God.

Straw skeps and a beekeeper. Woodcut engraving from Sebastian Munster's Cosmographia (Bern, 1545).

Medieval bees were generally kept in "skep hives," woven from straw coils with round conical tops smeared with cow dung as waterproofing. Skep hives were much lighter and easier to manage than the clumsy hollow log hives that they replaced. They were also inexpensive to make and could be expanded as the colony grew by adding more straw coils, called ekes, to the bottom (hence the expression "to eke out"). The downside of straw hives was that they sometimes went up in flames when their owners suspended them over fire pits in order to pacify the bees with smoke.

Not only did skep hives provide housing for bees throughout medieval Europe, they were also symbols adopted by the beekeeper and candlemaker guilds and appeared on signs over candle shops to advertise their wares. Skep hives are still considered the symbol of beekeeping and can be found imprinted on modern ceramic honey pots.

Forest Beekeeping

The notion of keeping bees in their original forest nests, from which beekeepers could harvest honey and beeswax at will, may have begun as an intermediate stage between honey hunting and true beekeeping in apiaries. There is archaeological evidence that forest beekeeping was practiced two thousand years ago in the heavily wooded areas of northern Europe, but we mainly know about it from

Eighteenth-century German forest beekeepers working their arboreal nests. From J. G. Krunitz (1774).

the charming woodcut prints in medieval manuscripts depicting forest beekeeping scenes.

Forest beekeeping was especially common in medieval Russia. The forests, most of which were owned by princes, boyars, and monasteries, were worked by beekeepers known as *bortniks* (*bort* means "hollow tree trunk"), who paid sizeable rents to their landlords. Bortniks cut distinguishing marks into the bark of bee trees to stake their claim to the residents and the honey they produced. In the eleventh and twelfth centuries, laws were passed to safeguard the bees by imposing heavy penalties on anyone found destroying a bee tree.

When forest beekeepers claimed a wild nest, they often enlarged the tree cavity and expanded the entrance in the front of the nest to make harvesting easier. If the nest was located high in a tree, the beekeepers cut footholds into the trunk to facilitate their ascent. Forest beekeepers also used climbing ropes and ladders to gain access to the treasure contained in the tree hollow. To discourage competition from honey-hungry bears, the beekeepers sometimes hammered sharp spikes into the tree trunk.

Forest beekeeping was also practiced in medieval Germany and Eastern Europe. In England, you can still see remains of the earthen embankments and stone walls that protected medieval bee gardens from the ravages of wild pigs, badgers, and other rapacious honey hunters.

With the coming of the Reformation, beekeeping in Europe, both forest and apiary, began to wane, just as it had after the fall of the Roman Empire. The Reformation resulted in the abolition of hundreds of monasteries and the dispersal of their beekeeping monks. The new Protestant churches, less focused on the cult of Mary and the virgin birth than the Catholics, had no need for expensive beeswax candles. The diminished use of beeswax occurred at about the same time that sweeteners other than honey became popular. Sugar was appearing on the scene, as was molasses, derived from sugar. Since it was cheaper to produce and transport than honey, sugar traders were able to undersell their beekeeping rivals—yet another blow to the honey industry.

Stingless Bees and Their Devoted Mayan Keepers

Among the world's beekeepers, none surpassed the pre-Columbian Maya in their devotion to their fuzzy, endearing little charges. The ancient Maya, who inhabited the region of what is today southern Mexico, lowland Guatemala, and central Belize, developed colorful rituals to define and celebrate their complex relationship with their bees. Over the course of many centuries, these elaborate rituals became one of the cornerstones of their remarkably enduring culture.

The husbandry of stingless bees (*Melipona* and less frequently *Trigona*) among the Maya dates back at least a thousand years. Not only did they cherish their bees—which nested in hollow logs in village gardens—but they depended on the honey the bees produced for a wide range of uses. It was considered an effective treatment for cataracts, conjunctivitis, and chills and fever. As an offering to the gods, it was believed to sweeten the disposition of even the most bloodthirsty deity. And of course it was a prized food. Honey from stingless bees had (and still has) a complex, distinctive taste, quite different from that produced by the more common European honey bee. Honey was also an important item of trade for the Maya, sent by sea to Honduras and Nicaragua as well as by land throughout Mexico. In exchange for honey and beeswax, the Mayans received cacao seeds and precious stones. Stingless bees were not only a critical source of medicine, food, and trade for the Maya, but also essential to their agriculture, for the prodigious bees pollinated no fewer than sixteen crops grown in the region. Early on, Mayan farmers had recognized the importance of stingless bees as pollinators and kept them in colonies near their doorway gardens. They also processed beeswax into molds for the lost-wax casting of gold jewelry and religious objects.

By the time of the Spanish conquest, stingless beekeeping was all but ubiquitous among the Maya, as can be gauged by looking at the official records of tribute paid to local governments. In 1549, nearly 95 percent of all villages in the Yucatán Peninsula paid a tax of honey and wax.

But bees and honey were more than just consumer items to the Maya. Along with other Mesoamerican peoples, the Maya mythologized the natural world in which they lived. The animals and plants they harvested from the wild or raised at home had religious and spiritual meaning as well as economic significance. It's not surprising, then, that brightly colored, highly stylized images of bees and bee gods adorned both the interior and exterior walls of many imposing temples.

We don't know exactly how the Maya became beekeepers. Perhaps in the early stages of their culture, when they were still hunter-

*A Mayan deity raises a log hive (*jobone*) of their sacred bee,* Melipona beecheii, *the prized honey producer of the Yucatán Peninsula. A stylized bee hovers nearby. From the ancient Mayan screenfold book, the Madrid Codix.*

gatherers, they witnessed giant anteaters and other mammals stealing honey from bee nests in hollow tree cavities, and realized that helping themselves to the sweet treasure inside would be a whole lot easier if they chopped the tree down first. And having cut down the trees, they may then have brought the sections where the bees nested back to their villages—a practice that would have led directly to beekeeping.

Although all of this is speculation, we do know that the Maya kept their bees in hollowed logs called *jobones,* each of which contained a nest that was two to four feet long. The logs were stacked on an A-frame rack that was kept in the shade of a thatched, palapa-like hut, called the *nahil-cab* or bee house. An entrance hole, often marked with a cross, was made in the middle of the log. Both ends were sealed with clay or stone discs to keep out ants and other predators. The beekeepers could simply remove an end plug and reach inside to harvest the honey. Many of these stone discs, elaborately carved with Mayan glyphs, have been recovered by archaeologists throughout the region.

Mayan beekeeping practices adhered strictly to traditions established in prehistoric times and recorded later in a thousand-year-old document called the Madrid Codix, found in the Yucatán by eighteenth-century Spanish explorers (its exact provenance remains unknown). The sacred text made it clear that the fates of bees and their human keepers were inextricably intertwined. When, for example, a beekeeper died, his heir had to inform the bees of the sad news while reassuring them that they would still be well cared for. The new beekeeper could not participate in any of the death rites lest his sadness disturb the sensitive bees. Should he visit a cemetery, the beekeeper could not interact with his bees for at least three weeks or serious harm would come to them. No Mayan would ever have asked a beekeeper to help lay out the corpse of a departed family member or friend, for if a beekeeper touched a dead body, he had to wash his hands and arms several times a day for three weeks with the leaves of an orange tree. He had to perform these cleansing rituals thoroughly before daring to approach his hives. If a bee was accidentally killed, it had to be carefully folded in a leaf and solemnly buried.

Carrying out these rituals was an intrinsic part of the Mayan beekeeper's life. If the rituals were not honored, the bees could become unproductive. Another form of insurance against failing productivity was the beekeeper's constant dialogue with his bees. Staying in touch with the bees helped keep their special relationship going.

Numerous findings from archaeological sites bear witness to the reverence the Maya felt not only for their stingless bees but also for Ah Mucen Cab, the god of beekeeping, bees, and honey. The Madrid Codix clearly directs beekeepers to honor the god with festivities to ensure a good flow of honey in the coming season. The many depictions of Ah Mucen Cab which have come down to us from those times are further evidence of the Maya's esteem for him. Ceramic incense burners from the Classic period of Mayan culture, AD 250 to 900, portray the god hanging upside down and glaring out at the world. (Unfortunately, scholars can't explain why the god is sometimes depicted in such an angry mood.) The Temple of the Descending God in the ancient Mayan city of Tulum is decorated with a carved relief showing Ah Mucen Cab holding a clutch of honey pots in his hands.

The U Hanli Cab: An Endangered Tradition

Scattered throughout the Yucatán and Central America are the remains of great Mayan cities whose temples, palaces, and pyramids were all mysteriously abandoned to the rapacious forests many centuries ago. This homeland of the ancient Maya is still inhabited by their descendants, living in impoverished villages where the first language is Yucatec Maya, not Spanish.

Today, visitors to Quintana Roo, along the southern coast of the Yucatán, travel on hot, crowded buses to see the ruins of Tulum, dramatically situated on a bluff overlooking the turquoise waters of the Caribbean Sea. This region was the last holdout of Mayan civilization, a remote area where small communities managed to survive with their culture intact for at least a century after the Spanish had subjugated the rest of the region.

The Yucatán is a broad, verdant peninsula thrusting eastward into the Caribbean and the Gulf of Mexico. When the Maya were still masters of this land, it was clothed in low-stature tropical forests that grew on a foundation of limestone, which was riddled with secret caves and underground waterways. Now, only a few patches of this lush forest remain, most of it having given way to cornfields, village gardens, sugarcane plantations, and beach resorts for visitors from the north.

Despite all the changes, if you ever visit the Yucatán, as I have done many times to observe what remains of the ancient beekeeping

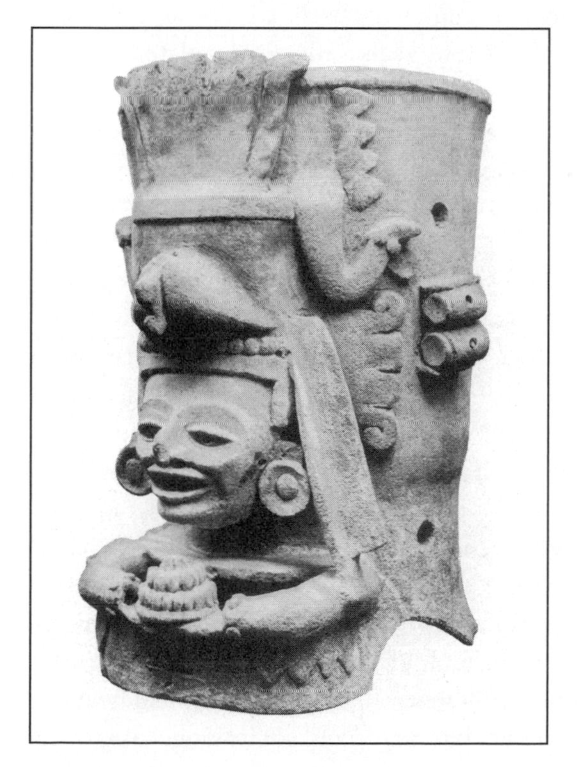

A ceramic incense burner from a Mayan temple in the Yucatán Peninsula depicting "Ah Mucen Cab" cradling a cluster of Melipona *honey pots in his hands.*

rituals, you may be surprised at how strong a hold Ah Mucen Cab still has over the Maya, at least in some of the more isolated villages of the countryside. The typical Mayan village of today tends to be a haphazard collection of thatched huts set amidst stately sabal palms, with an immense kapok tree lending its august dignity to an otherwise unremarkable village square. Flocks of noisy green parrots roost in nearby trees, loudly announcing the arrival of any outsiders. A few underfed dogs rouse themselves out of their torpor and amble over to sniff the visitor's ankles. The human inhabitants also rouse themselves, abandoning the shady hammocks where they have been taking their afternoon siestas, to find out what the visitor wants—which in my case is generally an interview with the local beekeeper, if there still is one.

It is in villages like these that an ancient beekeeping ceremony, the *u hanli cab,* is still performed, though what used to be an annual event now occurs only once every few years. Strictly regulated by immutable tradition, the purpose of the *u hanli cab* is to ensure healthy hives and bountiful honey harvests.

The preparation of special offerings to Ah Mucen Cab by the village shaman and his apprentices generally requires several days. These offerings include *balche* (a fermented honey wine flavored with the bark of the balche tree), thirteen hand-rolled cigarettes in cornhusk wrappers, strings of colorful glass beads, and a small gourd cup. The offerings are carefully arranged on a wooden table placed near the stacked log hives—thirteen cups aligned in three parallel rows, containing either *masa seca,* a corn flour gruel, or *balche.* A wooden crucifix dressed in brocaded vestments (a post-Conquest addition) is placed near the edge of the table, with the cornhusk cigarettes dangling from its arms. When the table has been properly set, the shaman sacrifices three hens, drowning each one by pouring four tinctures of *balche* into its beak with a folded leaf. Next, he produces two fresh turkey eggs, which he holds over the smoke of a burning copal incense pellet while he circles the table and chants a prayer. After the prayer, the chickens are butchered and cleaned. The menu for the festivities is dictated by ancient tradition, passed down orally from generation to generation. Many of the foods are made of *masa seca* that has been mixed with honey. Some of the *masa seca* is flattened into tortillas, which are cooked on a griddle, then sprinkled with parched squash seeds. The remaining *masa* is fashioned into large balls and wrapped in leaves. The *masa* balls and the chicken pieces are placed in a preheated roasting pit and allowed to cook for about two hours.

While the shaman carries out the various duties of his office, the village women prepare two kinds of soup, one red, the other white, and boil the smoked turkey eggs with the sweetbreads of the chickens. When the roasting pit is at last opened, the shaman sprinkles *balche* over the cooked food and chants a blessing. He then selects the choicest offerings, which are placed on top of the log hives under the shade of the thatched bee house. The cornhusk cigarettes are distributed to the revelers, who by now are well lubricated, having downed several shots of the potent *balche*. Finally, using a folded leaf, the shaman pours *balche* into the openings of the hives. When the ceremonies have been completed, the assembled participants help themselves to the remaining food.

Twenty-four hours after it has begun, the sacred *u hanli cab* is over. If he has found the offerings to his liking, Ah Mucen Cab can now be counted on to provide the bees with plenty of nectar and their dutiful keepers with bountiful harvests.

Because of the decline of traditional beliefs, the expense of the ceremony, and the advancing age of most of the shamans, the *u hanli cab* is now rarely performed. With the erosion of this tradition, the people are losing an important link to their own unique past. Cultural memory fades quickly in the absence of the ceremonies that embody it—a tragic loss both for the Mayans and for all of us who appreciate their ways.

A Visit to the Maya

Though the first European colonists in the Yucatán valued the wax made by stingless bees, they knew that European bees produced greater quantities of honey and, driven by a hunger for profit, began introducing them into Mexico in the sixteenth century. But more productive or not, foreign bees were rejected by the Maya, who felt that their native honey was far superior. For one thing, they believed it was more effective as a medicine for treating a variety of ailments. They also preferred its sweet yet slightly acidic flavor to the bland taste of the European honey, and they appreciated the fact that it was easier to store since, unlike non-native honey, it didn't crystallize. But the main reason they remained faithful to their own honey was the pivotal role it played in their religious ceremonies—a role no other honey could fill. And not only that, but their traditional beekeeping rituals, which ensured a productive relationship between the bees and their keepers, would be of no use in the management of European honey bees. If they accepted the bees being urged on them by the colonists, their crucial, ongoing communication with their own bees—a deeply caring dialogue, so alien to the Spanish that they considered it a kind of witchcraft—would fall into silence.

So the Maya stubbornly held out against the imported bees long after most of the New World had succumbed to the pressure to switch.

But now all that is changing. Not only are important Mayan traditions such as the *u hanli cab* on the verge of extinction, but deforestation has paved the way for large plantations, which use insecticides that have virtually eliminated stingless bees from many areas of the peninsula. In fact, the number of managed stingless bee colonies in the Yucatán has fallen to less than half of what it was in the 1980s. The gene pools of these bees are in steep decline, and so is the pool of traditional knowledge about their use and management. With the loss of stingless bees (recently estimated at 93 percent during the past thirteen years), many Mayan rituals in which honey and beeswax play an important part are now threatened with extinction—just as the Maya of 350 years ago had feared would happen.

If indigenous beekeeping is allowed to die out, another link between human society and the finite natural resources it must learn to manage will be gone. We feel that one way to prevent this from happening is to record the practices of traditional beekeeping before they are forgotten, in the hope that this will encourage the younger generation to carry them on. It was for this reason that my colleague and fellow entomologist Jim Donovan and I traveled to Quintana Roo in July of 2003. With the support of several U.S. foundations, we've launched an ambitious program to try to conserve Mayan beekeeping, while at the same time protecting the stingless bees themselves and their forest habitat. Armed with digital cameras, DAT recorders, and video camcorders to document the state of beekeeping in villages near the city of Felipe Carrillo Puerto, we pursued our long-term goal of creating a manual in English, Spanish, and Yucatec Maya to capture the ancient methods before they disappear forever. As soon as we have finished the necessary work and gotten the additional funding we need, we intend to distribute the manual to Mayan villages where this husbandry is still being practiced.

After a four-hour flight from Phoenix, our plane touched down at Cancún International Airport amid gray clouds and dramatic flashes of lightning. As soon as we had loaded our luggage and equipment into a rental car, we headed south to Akumal, a shallow coral lagoon where giant sea turtles lay their eggs on a pristine beach, not far from the impressive seaside ruins at Tulum. We made good time on Highway 307, and soon after arriving in Akumal and unpacking, we were pleasantly lulled to sleep by the sound of waves crashing on the shore.

The next morning we again headed south on Highway 307 toward

Belize. Although I'd been to the Yucatán many times before, both as a vacationing tourist and as a scientist investigating the local bees, I was unprepared for the incredible heat. It was mid-July, and the summer rains had already begun, causing both the humidity and the temperature to soar. After an hour and a half, we arrived at Felipe Carrillo Puerto, the historical and ceremonial heart of the region, where we met a Mexican colleague, Dr. Rogel Villanueva, a professor at El Colegio de la Frontera Sur (ECOSUR) in nearby Chetumal, on the border with Belize. With him were his two Mayan assistants, Margarito and Wilberto, who would be our interpreters during the coming days. Rogel is a palynologist, a scientist who studies pollen and larval feces to discover what kinds of flowers and plants the bees have visited. Working with Smithsonian's Dr. David Roubik, he unearthed evidence that the recently introduced Africanized honey bees are competing for food with the native stingless bees.

As we pored over detailed maps of Quintana Roo, Rogel showed us the sites he needed to visit for his pollen studies and pointed out villages where we would be able to witness the ancient *Melipona* beecraft. Many of his contacts had given up the stingless bees, either because their bees had absconded or died off or because they had switched to honey bee husbandry. Some, however, were still in business and during our visit, we planned to interview, photograph, and videotape as many of them as possible.

The following morning we headed north to the village of Tihosuco, past second-growth forests, roadside huts, cattle pastures, and well-tended fields of maize, beans, chilies, and various medicinal plants used by local Mayan healers. The massive ruins of the burned-out church in Tihosuco, standing on a hill overlooking the market, gave testament to the upheavals that had ravaged the region during the civil wars of the nineteenth century.

After we toured the small museum housing artifacts and photos of the period, Wilberto and Margarito asked the villagers if anyone still kept stingless bees in log hives. Soon we were on the trail of a local beekeeping family. When we arrived, two women invited us inside a small one-room house, furnished with sleeping hammocks that stretched from wall to wall, a couple of well-worn chairs, and a wobbly table. Several small children peered out at us from behind their mothers' voluminous skirts, watching the proceedings with wide-eyed curiosity. One of the women, resplendent in an embroidered blouse known as a *huipil,* went to a corner and brought back two big plastic Coca-Cola bottles filled with honey. Was it from the stingless bees? Rogel asked to sample it, and his smile gave me my answer. It was indeed, for it had the piquant aftertaste that we had come to know and appreciate, a far cry from clover honey squeezed out of a plastic bear. It had been harvested from colonies tended by the woman's husband, not twenty-five yards from where we stood. Was it for sale? To our surprise, the answer was yes. After some negotiations in Spanish and Yucatec Maya, a price was settled on. We could have both two-quart bottles for 400 pesos, about $20 U.S. It was a real bargain, for this kind of honey can fetch up to 300 pesos or more per quart. Usually one colony only produces one to one and a half quarts a year, but these colonies had produced two quarts each.

Now that business was settled, we were anxious to visit the bees. As we walked through the village, chickens, turkeys, and dogs scurried out of our way. Then it came into view: a magnificent pole-and-thatch bee house, with one colony of *Trigona* bees and twelve thriving *Melipona* colonies. We couldn't believe our luck.

The two women told us that the beekeeper's brother was a carpenter, which explained the fine craftsmanship of the bee house and its hollow-log hives. While Rogel busied himself with interviews, asking the women about the history and health of the colonies, how the beekeeper managed them, how much honey was produced, and where they sold it, I took dozens of digital photographs and shot some footage with my camcorder. Trying to be discreet, I videotaped a group of village kids sitting near the hives, then turned my attention to the comings and goings of the foraging bees as, one at a time, they scooted past the guard bees through the small round entrance holes into the colonies. The beekeeper's children watched us with interest and perhaps amusement, probably wondering what all the fuss was about. After an hour or so, we said good-bye to the family, assuring them that we would return one day to sample more of their delicious honey.

A few days later, we drove to the small Mayan village of Tuzik, where we met the wife of Don Ysidro Pedro Cruz. The *señora* was a small woman in her mid-sixties whose long, glossy black hair was threaded with a few strands of silver. She was wearing a typical Mayan dress of white cotton with colorful embroidery decorating the front.

She and Don Ysidro live in a traditional thatch-roofed hut set amid animal pens and gardens planted with beans, chilies, and squash. While waiting for her husband to return from a meeting in a nearby town, the *señora* talked with us about their bees. Don Ysidro used to have about forty *Melipona* colonies, she said, which he had inherited from his father, but many of them had aged and died out, and

Traditional palm thatch Mayan bee house. Log hives of Melipona *bees are stacked against upright poles.*

he now had only eight colonies left. He harvests about three pounds of honey from each colony once a year and sells much of it to the honey-packing cooperative in Felipe Carrillo Puerto. When Don Ysidro arrived, he lamented the loss of so many of his colonies but seemed resigned even as he shook his head and murmured about changing times. He was a compact man in his early seventies, with a black mustache and black hair. Like most village men, he was dressed in a plain white shirt, sandals, and a pair of well-worn trousers.

By now it was early afternoon, and we were ready to move on to see what kind of beekeeping we would discover in the next village down the road. But as we were saying good-bye to the old couple, I noticed the hatch marks that scarred the bark of the largest tree in the yard. Don Ysidro explained that they were there for a good reason, for this was a chicle tree. Don Ysidro, it turned out, is a part-time *chiclero,* extracting sap from the tree and producing chewing gum to sell in local markets. Tasteless and pink (Wrigley's had yet to make its sugary mark on this gum), it can be chewed forever without losing its resiliency.

Our trip to Quintana Roo confirmed our fear that the husbandry of stingless bees is rapidly dying out in the Yucatán Peninsula. In addition, colonies of stingless bees are harder to find in the wild, due to the decimation of so many forests for timber. During our interviews, we learned that while the present-day beekeepers' grandfathers and fathers had routinely made colony divides, thereby increasing their numbers, this is no longer being done. As a result, colonies are dwindling at an alarming rate. Moreover, as Rogel explained to us, *Melipona* and *Trigona* colonies are threatened by growing numbers of feral and managed Africanized honey bees. These introduced bees are taking the lion's share of floral nectar, making it difficult for native stingless bees to earn a living.

The problem isn't just confined to Quintana Roo. The truth is, honey bees of the genus *Apis* (which includes European honey bees) are marvelously well adapted and accepted—in their homelands, where they evolved millions of years ago. But when they are introduced to new habitats, they, like so many other invaders, pose problems for the locals because they don't quite fit in. As foragers par excellence, honey bees harvest so much pollen from the environments they have immigrated to that they literally steal food from the mouths of native bees and other pollinating insects, birds, and bats. For this reason, and despite the fact that they are such admirable creatures, we should be cautious in welcoming honey bees with open arms into all the habitats and landscapes of the world.

What impact will the disappearance of stingless bees and traditional beekeeping have on Mayan culture and the ecology of the Yucatán Peninsula? A huge one, in my opinion. The rituals attached to beekeeping are deeply embedded in the Maya's sense of themselves, and uprooting them will undermine the unique identity these people have preserved for so long and in the face of such overwhelming obstacles. Many of these rituals, as well as the deep understanding of the natural world they symbolize—an understanding that has evolved over many centuries—have regulated the way the Maya interact with and successfully manage their environment. Keeping the forest intact, refusing to use harmful pesticides, living within nature rather than trying to subdue it—these are the qualities that may well be lost with the demise of stingless beekeeping.

In the hope that we can help stem the tide of this loss, which has so many ramifications, Jim and I are now committed to working with Rogel and his Mayan beekeepers. Since returning to Tucson, we have been collating our digital photographs, editing our videos, and working on the trilingual manual, which will be illustrated with ancient Mayan bee glyphs. Our goal is to raise enough money to ensure that our educational materials will reach every Mayan village where the keeping of stingless bees, however endangered, still lives on.

Chapter 4

A Year in the Life of a Beekeeper

He climbed into the cab. The trailer was low, only a few inches above the wheels, and the mudflaps nearly reached the ground. The engine started and the cab shivered, but the load settled easily as it moved onto the highway. Semis were made for the big interstates, and the interstates for them. Andy Card rolled south with his bees, into the next season.

—Douglas Whynott, *Following the Bloom: Across America with the Migratory Beekeepers*

Today, few people go into beekeeping with dreams of becoming rich. Although there are some millionaires among the top American honey producers, for most beekeepers it's a matter of passion—quite simply, they love the bees they tend.

The Passion of the Beekeeper

My favorite time of year is the fleeting spring, which arrives in the Sonoran desert around my Tucson home in

mid- to late March and has departed by early April. Barely a desert, since it averages twelve inches of rain a year, the Sonoran is home to paloverde trees, shrubs, and annual wildflowers, which in spring paint the washes, hillsides, and creosote bush flats with their magnificent, saturated colors—the oranges of Arizona poppies, the cobalt blue of phacelia, the paler blues of lupines, and the pinks and reds of tall-stemmed penstemons and the shorter owl's clover. These flowering plants deck themselves out in such colorful finery to attract prospective sexual go-betweens, the bees, who, as dedicated pollinators, are necessary to their successful reproduction.

Spring is a time of rekindled memories for me, a spiritual and ecological awakening. I particularly appreciate the Sonoran spring because it's like no other season anyplace else on earth.

In the 1990s, when I kept three or four hives on my property, I loved to visit my bees during the cool morning hours of spring. Picking a spot a few feet behind and off to one side of the hives, and making myself comfortable on a flat rock or in a short-legged lawn chair if I remembered to bring one, I would watch thousands of female workers as they joined the morning rush hour to spend the day foraging for nectar and pollen. From my vantage point, their paper-thin, transparent wings dazzled in the clear morning light.

My awakening colonies all teemed with life as their hardworking citizens got down to the various tasks at hand. As one group of ten or fifteen bees departed, another returned, many of them bearing yellow, orange, and white pollen loads on their broad hind legs. The bees without colorful pollen loads were probably nectar gatherers. About three-quarters of the bees were after nectar, while the remaining fourth foraged for pollen to make into nutritious "bee bread." Usually a honey bee worker is either a nectar gatherer or a pollen forager. However, if a nectar gatherer spots a tempting bit of pollen, she may go for it and return to the colony well stocked with both types of food. Over the course of her life, a worker bee will alternate periods of specializing in nectar with periods of specializing in pollen.

To me, the nests seemed like giant pairs of lungs, inhaling the returning foragers, exhaling the departing ones. That was my favorite way of thinking about my hives—living organisms breathing out pollinators, who in turn breathe life into the desert plants and the entire ecosystem.

Nothing was more calming to me than escaping the phone calls, fax machines, and looming deadlines at my office in order to spend some quality time with my bees.

Like most beekeepers in the United States, I kept my charges in those familiar white wooden boxes named Langstroth hives after the man who invented them in the nineteenth century (see page 74 for a detailed description of the Langstroth hive and its history). The entrances to many of my hives were outfitted with pollen traps, an ingenious if somewhat malevolent device created by curious bee researchers. These traps taught me a lot about the ways of bees and their link to the flowering plants in their foraging domain of roughly five square miles. A pollen trap is made up of closely spaced but slightly askew screens. In order to enter the hive, incoming bees have

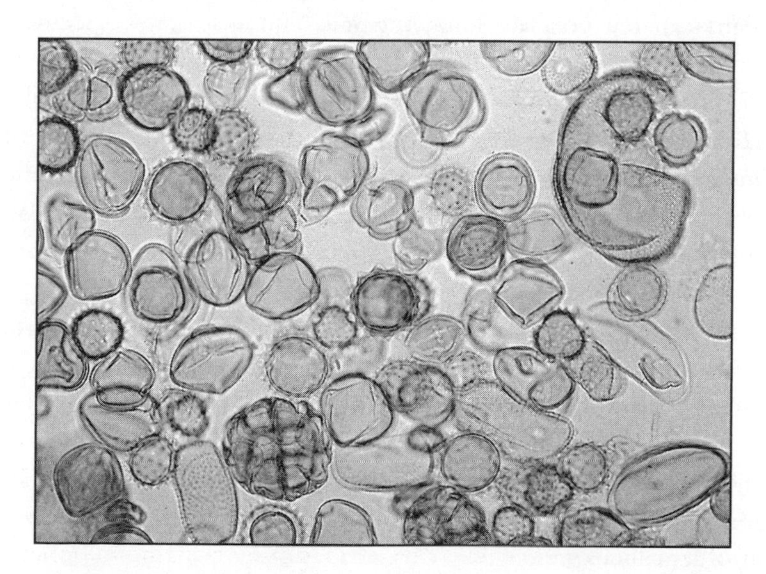

Typical pollen collected by bees in Arizona magnified by a light microscope.

to run the wire mesh gauntlet, losing about 65 percent of their pollen pellets, which fall into a drawer at the bottom of the hive. All I had to do was sidle up to a hive's backside, open the drawer, and take a sample of pollen. Reading the colors and microscopic identities of the grains told me what plants the bees had recently raided.

Sometimes I took fresh honey from a hive and studied the pollen grains it contained. For many years, I was a floral forensics investigator and was able to tell what families and genera of plants my bees had visited by literally looking into their pantries. The record of trace pollen grains in honey gives a reliable indication of its floral origins.

When honey is diluted with water, then spun in a high-speed laboratory centrifuge, the pollen separates out and forms sediment at the bottom of the centrifuge tube. After some chemical mischief with hot acids and a fume hood, the pollen is transferred onto microscope slides. Magnified four hundred times or more, the grains give up their secret identities, for each type of pollen is unique in its size, shape, and surface sculpturing. A good palynologist (a scientist who studies pollen) can tell you the family, genus, and sometimes species to which any pollen grain belongs. This is important information for beekeepers, since pure, single-source honeys command top dollar, making it necessary to know what kind of blossoms the honey has come from. By sending a sample to a laboratory and having its pollen content analyzed, beekeepers can find out if their tupelo honey really is from tupelo trees. Some countries require honey dealers to submit their unifloral honeys to a certified lab for quality control.

I always enjoyed removing the glistening honeycomb frames from my hives. As rivulets of honey trickled down the face of the comb, they were chased by bees trying to recapture the golden droplets. Ever eager to satisfy my sweet tooth, I would thrust a finger into the comb and lick it clean. Honey is at its best right out of the comb, still warm, and fresh and delicious in a way that is hard to imagine unless you've actually experienced it.

One of the pleasures of beekeeping is the wonderful fragrance of a hive. Once you've inhaled the aroma, it's something you'll never forget. Open a hive on a hot day, remove the cover and inner lid with your hive tool, and take a deep breath. The essence of the inner hive and its bee inhabitants is in that smell.

But the hive odor is difficult to place. Sometimes it's reminiscent of a brewery or winery, sometimes of a bakery filled with fresh pastries, especially if some of the combs are uncapped. There are also the rich aromas of beeswax and the medicinal scent of the plant resins that the bees turn into propolis, the glue that keeps the combs in place. Occasionally, a whiff of something unpleasant grabs your attention—the odor of dead and dying larvae afflicted with foulbrood, a bacterial disease that can ravage a colony.

During my beekeeping years, I was instinctively drawn to the sisterhood of the hive and never tired of peeking inside to see what my bees were up to, perhaps even to have a conversation with them, much as the Mayan beekeepers do. I also loved removing the brood frame and holding it up with my back to the sun so that the light penetrated its dark recesses, revealing the grubs curled inside their nurseries, maturing adults chewing their way out of their cells, and the queen herself slinking away from the light, to which she is unaccustomed, and trying to hide under the seething blanket of her daughters. All of these are numinous experiences, difficult to explain to a nonbeekeeper.

I don't, however, like getting stung, and I have never understood the machismo of some beekeepers who wear their multiple stings like badges of courage. (I know bee stings are supposed to relieve painful arthritis, but since I don't have arthritic joints, there is no point in enduring the stings.) To me, getting stung when working with bees means you are moving too fast, being careless and crushing your bees unnecessarily. A crushed bee gives off a bananalike scent from her mandibular glands, an olfactory alarm that triggers a stinging chain reaction in her sisters. As a result, the bees can go nuclear on you, trying to plant their barbs in your hide and willing to sacrifice their lives to defend the sanctity of their home.

From Hollow Logs to Man-Made Hives: The Story of Reverend L. L. Langstroth, the Langstroth Hive, and the Discovery of Bee Space

Many of us have seen apiaries, those collections of man-made hives arranged in neat rows along country roads, adjacent to blooming crops. But unless you are a beekeeper, you aren't likely to know how one man, the inventor of those boxes, changed the face not just of American beekeeping, but of beekeeping everywhere.

Lorenzo Lorraine Langstroth was a keen observer of honey bee behavior. An ordained minister, he remained a devout clergyman throughout his life. But he was equally devoted to his non-human flock—those winged, six-legged creatures who lived in special housing of his own design. Named after the inspired reverend, Langstroth hives are now used around the world by commercial and hobbyist beekeepers alike. For this reason, Langstroth is often regarded as both the father of modern beekeeping and the Henry Ford of hive technology.

Langstroth's insight about "bee space" revolutionized beekeeping. Bee space—or the distance between combs, about $\frac{1}{4}$ to $\frac{5}{16}$ inch—is the amount of wiggle room that bees need as they move around the dark confines of the nest. Bee space is the same whether it's in a man-made Langstroth hive or a hive of the bees' own construction in the wild.

In pre-Langstroth hives, there were no built-in supports for the honeycombs constructed by the bees. As a result, the bees built their combs wherever they pleased, and chaos and confusion reigned. Langstroth recognized that if he provided the bees with wooden frames on which they could hang their combs, and if he placed the frames at sufficient distance from each other to create ample maneuvering room, or bee space, the colony would develop in an orderly manner and thrive. Now, for the first time in history, bees were truly being managed. They could be made to build their combs where their keepers wanted.

Although the Langstroth hive harkens back to the Civil War, it has withstood the test of time and is used by most beekeepers.

Langstroth hives are typically made of pine boards half an inch thick. Their exterior surfaces are usually painted white to prevent the wood from rotting and to reflect sunlight in order to keep the hive a bit cooler. In early spring, the hive consists of two or three stories, which beekeepers call supers. Each super is a box that contains several removable rectangular wooden frames on which the bees can build their waxen combs. The frames hang vertically from top bars, inner rims along the upper edges of opposite sides of the super.

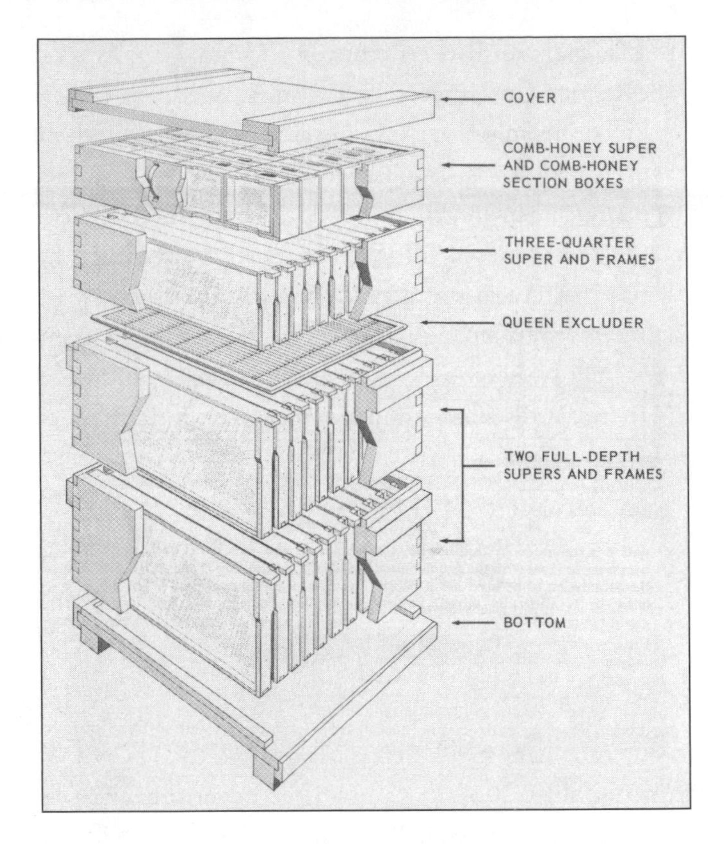

Langstroth-style honey bee hive.

The combs that are used for storing honey are in the shallower, upper supers, while those for raising the young are in the deeper supers below. As the seasons progress, the hive can grow twelve stories tall as beekeepers keep adding honey-collection supers to the top of the stack. Giving honey bees more storage room is a way to exploit their hoarding tendencies. The more room they have, the bigger the harvest.

The diagram gives us a look into the interior of a Langstroth hive with all its parts labeled. Unless you come from a beekeeping family, the terminology can become confusing. The supers rest on a bottom board that has wooden cleats nailed to its underside to give the hive a stable foundation on which to rest. Topping the uppermost supers, where the honey is stored, is the hive cover or lid. This keeps out the sun, wind, and rain and provides the bees with a cozy, comfortable home.

Let's examine the lowermost super of the hive, which rests on the bottom board. The bees build combs in the frames of this super, which they use as nurseries to rear their brood or as pantries to store the pollen they have made into bee bread. Little or no honey is stored here.

Resting on the bottom one or two supers is a metal grid that looks a bit like a barbecue grate. This is the queen excluder, a simple device that beekeepers use to manage their colonies. Worker bees can easily squeeze through the parallel bars of the excluder, but the larger queen cannot. With an excluder in place, the queen is confined to the nursery below, which means she can't lay her eggs in the upper supers, contaminating the pure honey that is stored there.

The upper supers are generally shallower than the supers below. This is because shallow supers full of stored honey weigh about thirty-eight pounds, easier to lift than the eighty pounds a deep super full of honey would weigh. Nevertheless, many a beekeeper complains of a bad back due to a lifetime of lifting and stacking full honey supers.

To avoid crushing my bees, I always worked with them in slow motion, paying careful attention to where I placed the heavy supers and gently nudging bees out of the way. A hive that is disturbed frequently, with too many unnecessary inspections and smokings, will produce less honey than one that is left at peace. A beekeeper should manage his or her colonies with gentle understanding and a thoughtful touch. Keep your populations up and give your bees plenty of room to produce honey and make babies. Remember, the great thing about these pollinating pets is that they do all the really hard work, collecting the nectar, ripening the honey, and packing it away in the combs. All you have to do is just let them bee.

The Beekeeper's Bees

While stingless bees have long been established in the Americas, honey bees, native to the Old World, didn't arrive until the sixteenth century, when Spanish settlers imported them, hoping to replace the bees of the Maya with their own, more productive kind. Soon the newcomers were swarming their way north toward what is now the United States, where they joined honey bees introduced in the seventeenth century by English colonists with beekeeping ambitions. Early accounts indicate that these enterprising bees, called "white man's flies" by Native Americans, were colonizing new territory at the rate of at least fifty miles a year. According to records of the period, the advancing honey bee front was often one hundred to two hundred miles ahead of the American frontier.

It was the hardworking honey bee that would eventually staff the Langstroth hives of most of America's modern beekeepers.

The Beekeeper's Year

Unless you buy your honey from a farmers' market, roadside stand, or beekeeping neighbor, it most likely comes from one of the huge honey marketers such as Sue Bee, a cooperative association of 375 large-scale beekeeping operations, which produces about forty million pounds of honey a year.

The big packers and distributors get their honey from beekeepers all over the United States, especially Florida and the Dakotas, bargaining with them to acquire their harvest at the lowest possible price. The largest beekeeping operation in the world is Adee Honey Farms in Bruce, South Dakota, with over seventy-five thousand hives.

Whether the colonies number ten or seventy-five thousand, there is a rhythm to the beekeeper's life. At certain times of the year, certain things need to happen in the beeyard, following a natural sequence of events that cannot be ignored or disrupted. These are the seasons of the beekeeper, which vary little, especially if the bees are kept in the temperate United States.

The Beekeeper's Spring

When the golden sun has driven winter
under the earth, and unlocked the heavens with light,
from the first they wander through glades and forests,
grazing the bright flowers, and sipping the surface of the streams.

—Virgil, *Georgics,* Book IV

Spring is the busiest time of year for honey bees and their keepers, whether the operation is in the desert uplands of southern Arizona, the citrus groves of Florida, or the apple orchards of Washington State.

The first wildflowers of spring beckon both the bees and their human landlords. In the far northern states, the bees have been hive-bound for months and are eager to get outside and buzz through the warm, fragrant air. Bees of foraging age act as scouts, the first to venture out into the big world beyond the hive. As soon as they locate the early spring blooms, they return to the hive and perform a waggle dance, which communicates the location of the new food sources to their fellow foragers. There may still be some fall honey stored in the hive, but there is probably very little pollen, or bee bread, left in the darkened brood comb, so the bees need to bring in lots of protein- and lipid-rich pollen pellets. The nitrogen and amino acids in the pollen will nourish new bees, which, when mature, will bring in yet more pollen and nectar to feed and fuel the ever-growing colony.

Good beekeepers inspect their colonies regularly during the spring. They know that if a colony is in good health in early spring, it will have stored a surplus of honey by late spring and early summer. (If they rent their colonies for crop pollination, a healthy population is essential.) The surplus honey will be harvested and sold at farmstands or to the big packers for distribution across the country and around the world.

Using the bellows of a bee smoker to puff cool smoke into the hive to quiet its residents, the keeper examines the lowermost super to gauge the condition of the brood, assess the health of the queen, and detect the presence of eggs, disease, or parasitic mites. In the uppermost supers of the hive, the beekeeper looks for evidence of new honey production: are the bees producing fresh wax to construct extra storage combs for their spring or summer crop?

Spring is the time to treat the bees for diseases such as American or European foulbrood, nosema, or chalkbrood. Healthy bees and a rapidly growing population are necessary for the colony to produce enough foragers to bring home the nectar and pollen. A weakened or sick colony is a drain on the beekeeper's time and finances. Weak colonies often require supplemental feeding with a sugar solution or corn syrup to grow rapidly and become strong enough to earn their keep. Sometimes a beekeeper will construct a sugar water feeder out among the colonies. There is the danger, however, that this open source of free sugar will incite a feeding frenzy, causing the bees to rob neighboring colonies to satisfy their craving for more carbohydrates. The bees can get ornery during a feeding frenzy, and beekeepers, friends, and neighbors are likely to get stung.

Spring, of course, also means spring cleaning—and the hives are in serious need of it. Mice may have entered some of the colonies in the fall and caused damage to the combs. These combs, as well as any damaged by tunneling wax moth larvae, are removed by the beekeeper and replaced with fresh or recycled frames. To allow the colonies to grow, the beekeeper may add an extra brood super holding eight to ten deep frames. Up in the honey attic, one or more shallow supers may be added to make room for the expected production.

The Beekeeper's Summer

A swarm of bees in May
Is worth a cow and bottle of hay
A swarm of bees in July
Is not worth a fly.

—"A Reformed Commonwealth of Bees," 1655

For most beekeepers, no matter where they live in the continental United States, summer, like spring, is a busy time of year. There's always something to do in the beeyard. What with harvesting honey, installing new supers, painting hives, fixing old bee trucks, and repairing honey house equipment, there aren't many moments when beekeepers are idle.

Collecting the nectar when it is flowing is the name of the beekeeping game. Although nectar flows can occur in the spring and even into the fall, bees stockpile their largest surpluses during the summer months. This is also the time of the year when honey bee populations are at their peak. Sixty thousand bees living in one hive box isn't uncommon.

During the summer, the bees work from dawn to dusk, each bee of foraging age making five, eight, or even twelve trips a day several miles from the nest. Sunshine is plentiful, grasses shimmer in the warm breezes, and flowers abound. This is a great time to be alive if you're a honey bee—or a beekeeper with many healthy colonies.

When the bees work from dawn to dusk, so do their keepers. If the nectar is coming in fast and furious (averaging three to five pounds per colony per day), the beekeeper must work hard assembling extra honey supers from parts previously ordered. As soon as the old supers are full, the beekeeper needs to remove them and replace them with the new ones, ready to take on more liquid gold. The secret of good beekeeping is knowing when the nectar flow is on and then gently managing the colonies by adding or removing supers and frames as required.

Summer nectar flows are the prime flows in almost all parts of the country. In my home state of Arizona, this is when the velvet mesquite produces a second flush of blooms and the groves gush with nectar and hum with bees in the branches overhead. Around Tucson, summer is the season when our state flower, the saguaro cactus, puts out its massive white blooms. The morning after the cactuses bloom, honey bees head for the nectar- and pollen-rich saguaro flowers, competing with native cactus bees, flickers, woodpeckers, and white-winged doves for the tasty treats. We also get good summer nectar flows from acacias, especially if the monsoon rains traveling north out of Mexico reach us during July and the plants respond by flowering early.

The summer nectar flow peaks at different times in different regions of the country. Some flows are brief, lasting but a day or a week at most. Other flows might go on for a month.

For the nation's thousand or so migratory beekeepers, summer's blooms are the signal to hit the road with their eighteen-wheelers packed with hives. Few people know about these relatively small-scale operations, which roll down the interstates, following the nectar flow for thousands of miles. When the mobile beekeepers find a promising spot, near an apple orchard or citrus grove, they unload their hives with small forklifts, and gently place them near the blooms so that the gathering of nectar can begin. The goal is to harvest a lot of honey, especially high-value, single-source honeys such as that from white clover and aromatic orange blossoms.

Many migratory beekeepers also derive a large portion of their income from pollination fees paid by farmers and orchard owners, eager to ensure that their fields and groves will be well pollinated in order to bear the highest-quality fruit. There are many long-established migration routes. The Dakotas, with miles of roadside fields thick with clover, the top-producing honey crop in the nation, have always been a mecca for migratory beekeepers during the summer months. Beekeepers in Florida follow the nectar flow northward along the eastern seaboard or head straight for the nectar-rich blueberry barrens of Maine. Beekeepers in the heartland go west to pollinate alfalfa fields and almond groves or to be in the right place at the right time for transient wildflower blooms.

In the past, beekeepers traveled by buckboard wagons or on boats and trains to follow "hot spot" blooms. A Chicago honey dealer by the name of Perrine bought a steamboat to ferry his thousand hives up the Mississippi River in the 1870s. Perrine and his crew set out from New Orleans in the spring, often anchoring near the fields of wildflowers that grew along the riverbank so the bees could gorge on nectar and pollen. Unfortunately, many of the bees took off and neglected to return to their floating homes, fatally compromising the venture.

In the early twentieth century, East Coast beekeeper Nephi Miller, reputed to be the first to produce a million pounds of honey in a single year, sent his charges by rail to winter in sunny California. Moving bees by train, however, was not very popular with the non-beekeeping world. The railroad workers were afraid of being stung. The horses that pulled the bee wagons to and from the railway depots were also less than thrilled, since bees and horses don't mix—many nervous steeds were stung and bolted away. And finally, the bees themselves had issues with riding the rails. The cars they were packed in were sometimes forgotten on side railings, where the bees literally cooked in the hot sun and the wax melted out of their hives.

In the roaring twenties, flatbed trucks were used to move not only moonshine but also honey bees. These dependable trucks could be easily loaded and driven to distant nectar flows. And best of all, they didn't complain about stings.

The Buzz at the Post Office

Early summer is the time for a beekeeper to expand the apiary by establishing new colonies. All the beekeeper has to do is buy two- or three-pound "packages" of bees from bee breeders and install them in a new hive box. Just add a similarly purchased queen (a bargain at $25 or less) and you have another colony, ready to produce. In rural post offices around the country, the arrival of a loudly humming package has caused more than one alarmed postmaster to place an urgent call to the beekeeper to collect the hive's new employees as quickly as possible.

The beekeeper mustn't procrastinate if there's to be any profit from these efforts. By midsummer, it's too late for a newly installed colony to take hold and build up a strong population. Without a lot of bees in the box, there aren't enough foragers to collect the nectar or enough honey makers to ripen the honey and store it in the combs.

Late summer is the time to collect honey produced during the nectar flows earlier in the season. Small-time beekeepers generally do their honey processing in their kitchens or garages. But large-scale commercial beekeepers load the honey-heavy supers from the hives onto a truck and transport them to a honey house, where the crop will be harvested.

The honey house usually consists of one room, from five hundred to a thousand square feet in size, with electricity, running water, and a smooth cement floor. The scent of hot honey and beeswax fills the air like sweet incense. Elevated temperatures keep the honey flowing, while heated electric knives make short work of uncapping the combs. (Hitam Manis would not be pleased with the use of a metal knife, electric or not.) In this mechanized age, a centrifugal honey extractor is used to get the honey out of the combs. (Would our Malaysian honey hunters be envious or scandalized at such a breach of tradition?) The supers are loaded onto the extractor and spun for several minutes as the strong centrifugal force sucks the honey out of the cells and throws it against the sides of the extractor; it then flows to the bottom. Next the honey is drained out of the extractor and pumped through hoses into settling tanks. When it has settled, it is packed into massive fifty-five-gallon drums and shipped off to the big cooperatives (such as Sue Bee) or to custom honey packers, depending on who is offering the best prices at the time.

You can usually locate the honey house in an apiary because it is garlanded with hundreds of bees buzzing excitedly around the building in a determined, if futile, effort to break in for a free meal.

The Beekeeper's Autumn

As summer advances into fall, there is a crisp chill in the air that migratory birds and busy honey bees notice long before the first frost warnings appear in the local papers. And as the season changes, so do the colors of the floral landscape. Pinks, blues, and lavenders give way to the warm yellows of the sunflower family. Asters, coneflowers, and goldenrod abound, spreading across meadows and bordering country roads and forest paths. This is the last big bloom of the year, and the pollinators know it. Honey bees, bumblebees, digger bees, and sweat bees all jostle one another on crowded blossoms, scrambling to harvest the last of the season's nectar and pollen. The race is on, for the winter ahead will be long and arduous, and it is critical to lay up enough stores to wait out the cold. There will be no flowers on the horizon or excursions out of the hive until the seasons turn and spring comes around once again. But there may be sufficient quantities of honey produced during this season to give the beekeeper one last harvest, while still leaving enough to get the bees through the coming months.

Beekeepers in autumn are just as busy as their bees. They must be especially vigilant, for though fall swarming isn't common, it does happen. Loss of half the population of a hive in the fall means the colony will probably die out over the winter. To prevent swarming, the beekeeper needs to add an empty super to the hive, filled with ten frames of beeswax foundation, so the colony has room to grow and the bees don't feel pressed to find a new home. Sheets of the man-made, hexagonally stamped foundation, consisting of pure beeswax sometimes coated in plastic, are purchased from a bee supply house and give the bees a head start in their construction efforts. Fall is also the time to combine two weak colonies into one strong one that is sure to last the winter and be ready for the spring nectar flow.

When fall comes, many beekeepers install entrance reducers in the narrow, mouthlike colony doorways. An entrance reducer is a thin piece of wood cut to the width and length of the entrance, with a notch in the middle to allow the bees to come and go. It fits snugly into the entrance opening and can be tightly secured with a nail or two. Entrance reducers keep winter drafts out of the colony and make it easier to prevent the invasion of field mice and other predators. Seeking the warmth and rich store of food found in the hive during the winter months, the mice can be very destructive to bee colonies. Although the bees eventually sting them to death and entomb them behind walls of the resinous bee glue called propolis, they can wreak considerable havoc before being dispatched to their untimely end. (Embalmed mouse mummies are an interesting side note to beekeeping.)

One disadvantage of entrance reducers is that they can become blocked with dead bees over the winter, making it difficult for the hive to do its spring cleaning. Some beekeepers don't use entrance reducers at all, figuring they'll put up with the mouse damage in exchange for better ventilation and hygiene.

Autumn can be a time of great floral abundance. If the colonies are located near a big meadow of goldenrod and asters, the bees may produce quite a haul. (Honey from asters is not everyone's favorite—many even consider it downright rank. But to me it's full-bodied and flavorful, capturing the true essence of the season.) Beekeepers who harvest honey in the fall must not be greedy, for if they take too much, the colony will run through its reserves before winter has ended and another round of nectar collecting and honey making can begin.

Fall is a good time to inspect the colonies. Are the bee populations strong and healthy, free of foulbrood, nosema, or chalkbrood? Has dry rot attacked the wooden hives, requiring new parts? Some beekeepers move their hives to sheltered spots out of the wind and protected from the cold. Winter is coming—the bees know it, and the beekeepers make ready.

The Winter of the Beekeeper

Winter, not surprisingly, is the time for "wintering" the hives to protect them from the change in weather. When I was keeping bees, I was fortunate not to have to do any wintering, other than making sure my colonies had adequate honey stores. For beekeepers at higher elevations or in the Northeast, it's an entirely different story. There, winter blizzards and intense cold pose a serious health risk for the bees. The thin pine lumber used in most man-made hives doesn't offer enough protection from the howling winds of a nor'easter. In these areas, the more thermal protection a colony gets, the less honey the hunkered-down bees need to keep their internal fires stoked and their cellular machinery going. For cold-weather beekeepers, wintering hives often involves wrapping the entire colony with insulating materials such as straw, plastic, or even Styrofoam.

Another threat during the winter months are the blankets of snow that can completely bury a colony. When this happens, the beekeeper has to brave the elements in order to clear the hive entrances so the bees inside won't suffocate.

Throughout the long, cold months, the bees congregate in what is called the winter cluster, a tight sphere of bee bodies forty thousand strong, usually located near their stored cache of honey. They are literally huddling to keep warm. It's called thermoregulation, and the bees are experts at it. By eating honey, then shivering their flight muscles without moving their wings, they can raise their internal body temperatures significantly. Revving their mini-engines keeps not only individual bees warm, but their neighbors as well. The temperature will not dip below 68°F within the cluster. When bees in the outermost layers start feeling chilly, they push their way deep into the center, the warmest part of the cluster. Wouldn't you?

With their clustering ways and fuzzy coats, the bees can survive freezing or even subfreezing weather. They do pay a price, however, for all those burned calories must come from their store of honey. If the beekeeper has taken too much honey in the fall, it could mean disaster in the months that follow.

Winter is a good time for the beekeeper to work indoors, boning up on the craft by reading *The ABC and XYZ of Bee Culture* or going through those back issues of *American Bee Journal* that there was no time for during the busier seasons of the year. Newer beekeepers might use the downtime to assemble more frames or repair damaged supers, hive lids, and bottom boards. A forward-thinking beekeeper will go online and look for the latest in disease-resistant queens from bee producers in Hawaii, New Zealand, or Georgia.

Despite the cold winds that blow regularly across the apiaries, there are usually a few warm days when temperatures rise above 55°F and the bees can escape their winter clusters and venture outside. There are no flowers to tempt them, but necessity calls. The problem with clustering is that the bees don't have an opportunity to relieve themselves. Since they're very hygienic creatures and will not foul their nests with their own excrement, they need to fly out of the hive to take care of important business. Beekeepers euphemistically refer to these winter potty breaks as "cleansing flights."

And now it's time for yet another change in season. As a result of all the beekeeper's cold-weather chores, the colonies come out of the winter months healthy, well fed, and ready to gorge on the abundant nectar and pollen of spring.

And Here Comes the Swarm

Swarming, a natural event in the biology of honey bees, is their way of increasing their numbers. It can happen anytime during the year, depending on the type of bee, the weather, and the flowering calendar in the area where the bees live. European honey bees, for example, usually swarm once or twice during the late spring, while the Africanized bees in southern Arizona swarm many times throughout the year.

The bees know when their colony has become overpopulated, whether it's established in the cozy hollow of a stately tree or in a manufactured wooden hive. Overcrowding is the signal to head out and start a new colony with room to grow and expand.

Prior to the actual swarming event, scout bees venture out on reconnaissance flights and inspect every hollow tree, rock outcropping, and large cavity in the neighborhood. Like any prospective homebuyer, they thoroughly inspect the premises. Is it dry inside, protected from the elements, and near plenty of flowering plants? Are neighboring (and competing) bee colonies too close for comfort?

Eventually, dozens or even hundreds of scouts converge on one spot. Somehow a consensus has been reached and group action takes hold. Now it's time to begin the move. Returning to the hive, the scouts perform waggle dances, thought to inform the other bees of the location of their new home.

Meanwhile, the scouts who have remained at the new home site to stake their claim to it form a circle around the entrance, facing outward. Assuming a characteristic tail-up, head-down stance, each bee fans its wings into a blur as it exposes the glistening Nasanov gland near the tip of the abdomen. The gland secretes a flowery potpourri of chemicals that will help the migrants find their way to the new residence.

As the swarming process begins, thousands of bees pour out of the old colony and settle on the ground. Thousands more clamber up the walls of the hive in dark festoons. Many take to the air and swirl in looping flights around the soon-to-be-abandoned nest. Eventually, the old queen, a few hundred drones (male bees), and usually half of the nest's workers (sterile females) are hovering in the air, ready to move to the new location. There are no real leaders, but somehow the system works and the swarm takes off, slowly at first, then faster, a giant, seething mass moving through the spring morning like an ominous storm cloud. The swarm may contain twenty thousand bees or more and can be one hundred feet wide and twenty to thirty feet tall.

I've been lucky enough in my career as a bee researcher and part-time beekeeper to witness the swarming spectacle dozens of times. I've even experienced the adrenaline rush of running *inside* several swarms as they traveled to their new lodgings. It's called swarm running, and I do it just for fun. The bees are gentle, their stomachs full of honey packed for the trip, and they are not in the mood to sting. As I run, bees swirl about me in all directions, but somehow the mass stays coherent, changing shape but not dispersing. As the swarm pulses its way forward, it ramps up to top "bee speed," 15 mph on the wing. Though a former marathon and cross-country runner, I have to struggle to keep up, and the swarm inevitably leaves me in the dust, my heart pounding as I gasp for breath and wonder where the bees will finally settle in.

Though I find running with the bees an exhilarating experience, to beekeepers swarms are nothing but trouble. When the bees leave the apiaries, they are usually gone for good. Each colony that produces a swarm loses half its bees and therefore half its honey-making potential. To prevent swarming and increase their colonies, many beekeepers make "divides." After selecting a large, populous hive, the beekeeper turns it into two hives by dividing the supers and placing them on two separate stands, each with bottom boards and a cover.

At least one or two frames of emerging or capped brood cells are placed in the lower supers, along with three or four frames covered with worker bees. Usually, no effort is made to locate the original queen. The beekeeper leaves it to Mother Nature and the colony to produce a virgin queen in the queenless half of the divide.

Chapter 5

Secrets of the Bee

Some have said that a share of divine intelligence is in bees.

—Virgil, *Georgics,* Book IV

The Sexual Dance of Flowers and Bees

Though we joke about the birds and the bees, we know that in fact there is an intimate association between flowers and the bees who play an essential role in their reproduction. We remember the idle summer days of our youth, following the zigzagging flight of a fat, colorful bee as it buzzed from blossom to blossom. The bee was probably a honey bee, genus *Apis,* of tropical origin. It sips the watery nectar that flowers secrete to attract pollinating insects. The nectar is a sugary bribe since the pollinators are trading sexual favors (giving a free ride to pollen grains, the male sex cells of flowering plants) in return for a drop of nectar and the chance to collect the protein- and lipid-rich pollen grains. Back at the nest, the bees concentrate the nectar, adding something of themselves, and store it in capped hexagonal combs made of beeswax. Once ingested, the sugars in nectar and honey fuel their flight and, as the nutrients are metabolized, keep the colony warm during long, temperate zone winters.

To experience a day in the life of a bee, we will take an incredible journey—seeing the world as the bees see it and learning the secrets of their highly sexual relationships with the many flowers they visit in the course of a busy afternoon. We will buzz through the air of a hot summer day, looking down on a world made up of vibrant colors and irresistible odors and shapes: a field of wildflowers, a suburban backyard garden, or a desert in full bloom. It's an elaborate mating ritual—and a desperate game of survival, for without the bees, the flowering plants could not procreate, and without the flowers, the bees would starve, their young would be stillborn, and their waxen nests would shrivel and fall into ruin. Think of it as an age-old minuet, the flower gently swaying in the breeze, the bee circling in for a landing to gorge on nectar and pollen, then, coy mistress that she is, taking off for another floral rendezvous. But the flower is satisfied, for as brief as the encounter may have been, it will bear fruit as the bee flies away with a film of fine golden dust on her knees and wings—life-giving pollen that will be well distributed throughout the course of the day as she keeps her many amorous appointments.

Sex Parts: That Flower's Packing a Pistil

Flowers are living billboards, vying for attention. Our planet is blessed with more than a quarter million species of angiosperms, or flowering plants. Their blossoms scream out in myriad sizes and shapes, exude nutritious oils and sweet, exotic, sometimes surprising scents, and display a dazzling array of color, the most saturated in all nature. They brought life to our largely green and brown landscapes long before we were there to stand up and take notice of them.

Flowers grow everywhere, from the frozen wastes of the Arctic to rocky desert outcrops and luxuriant rainforest canopies. They are so common we often take them for granted. Some, such as yellow dandelions popping up in manicured green lawns, we think of as weeds and relentlessly strike down with chemicals and spades. But where would we be without public and private gardens, prom corsages and bridal bouquets, the floral centerpieces that enliven holiday tables, alpine meadows blanketed with wildflowers, and of course perfume?

Though flowers have the power to seduce us, we are irrelevant as far as they are concerned, for their beauty, form, and function are not meant for us at all. Flowers evolved more than one hundred million years ago and have survived by partnering with creatures other than humans to propagate their species. It's all an elaborate game of sexual intrigue. They boldly advertise their wares, fine-tuned by natural selection with all the right enticements to catch the faceted, compound eyes of bees and other insects and to tempt the sweet tooths of birds, bats, and other mammals. Floral signposts beckon with all the colors of the rainbow and some that are beyond the rainbow, splashed across petals in ultraviolet wavelengths that only an insect can see.

The flowers give their pollinators what they want and are paid back with sexual favors. This shocking secret was discovered in 1750 by Arthur Dobbs, spying on the tulips in his back garden. Flowers are sex organs, plain and simple, but sex organs with a big problem. The male and female parts of flowers, the pollen and ovules, need to get together for procreation to take place. If the male gametes, inside the pollen grains, couldn't fertilize the ovules, destined to become seeds, there would be no more flowering plants. And without flowers, the world would be a very impoverished place. Being rooted to the ground, however, makes finding a suitable mate difficult. Flowers are all dressed up with nowhere to go. Like perfumed courtesans, they have to wait patiently for their partners to arrive. Luckily, there are sexual go-betweens, willing to transport the pollen from flower to flower. If you suffer from hay fever, you already know that one of these go-betweens is the wind, carrying pollen through the air, perhaps toward a receptive floral target. (It's more likely to find your nose.) It's all a bit hit or miss, not unlike the lottery.

The odds of making a successful match increase when the go-between is an animal or insect. About 80 percent of the world's angiosperms have their sexual gametes delivered by these messengers.

Among them, bees are the champions. They have more structural and behavioral adaptations for pollinating plants than any other creature.

Pollination: A Lucky Accident

Pollination, the transfer of pollen grains from the male part of one flower (the anther) to the receptive female part of another (the stigma), is really nothing more than a lucky accident—lucky for us and the millions of animals with whom we share the planet, since we depend on it for much of our food supply, and lucky for the flowering plants, which require pollination to reproduce.

Pollination is an accident because animals and insects don't set out to do their daily good deed by spreading the pollen around. They are just looking for a meal. For bees, pollen grains aren't the male gametes of flowering plants, but protein- and lipid-rich food that they depend on to nourish themselves and their hungry brood back at the nest.

Pollen grains are small and sticky, and they come in all shapes and sizes. Most have oily surfaces, and many have spines. When bees zero in on a flower, pollen sticks to the feathery hairs densely covering their bodies. Electrostatic charges make the seal between bee and pollen all the tighter, because the pollen grains are negatively charged while bees acquire a strong positive charge during flight. Besides the pollen they inadvertently gather while buzzing about in the innards of a flower—the pollen that will fertilize other flowers—the bees also harvest pollen for their own purposes. Some of them—honey bees and bumblebees—moisten the pollen they harvest with saliva, working it into a pliable mass attached to the smooth concave regions on their hind legs, the corbicula or pollen baskets. Other bees carry pollen back to their nests dry, transported among the coarse mats of thick hair on their hind legs. *Diadasia enavata,* a sunflower-loving bee of the United States, is so laden with bright orange pollen by the time it finishes its day's work that it looks like a flying Cheeto as it heads home.

What's in It for Them: Floral Rewards That Keep the Bees Coming Back for More

The flowering plants use bees and other pollinating animals as sexual go-betweens to get their ovules fertilized, creating the seeds that germinate new plants. As for the bees, they have a multitude of uses for the pollen and nectar they harvest from the flowers, ensuring that they will always be back for more. Pollen is the bees' primary source of protein, nitrogen, amino acids, and fat. Think of it as the beefsteak of the bees' diet. (While floral pollen is the perfect food for bees and certain wasps, for humans pollen gathered from honey bees and sold in tablet form at health food stores is not the miracle it's often touted to be. In fact, it's a very expensive way to acquire protein—and it's potentially risky, since up to 30 percent of the pollen collected by honey bees comes from nasty allergens such as Bermuda grass and ragweed.) Nectar is also a crucial item on the bees' menu, especially for honey bees, who use it as a side dish for developing larvae, and as fuel, powering the flight muscles of adult bees as they visit their floral

European honey bees (Apis mellifera ligustica) *on a dandelion.*

partners. In fact, honey bees are champion sugar junkies. The nectar they imbibe is from 30 to 50 percent sugar. Compare that to Coca-Cola, which is only about 10 percent sugar.

The flowers can't afford to give up all the pollen they produce as bee fodder. A few precious grains must make it to their respective stigmatic targets or the whole process will be fatally compromised. Obviously, both bees and flowers are getting what they need from the interchange, for the system of floral enticements has worked for millions of years, keeping our flowers fertile, our bees well fed—and our planet alive.

Pollen and nectar aren't the only bribes flowers have to offer. In the desert around Tucson, Arizona, plants known as malpighias and rhattanys have evolved specialized, oil-filled blisters on their outer surfaces. This edible vegetable oil is rich in energy and eagerly sought out by the fast-flying, aerobatic bees of the genus *Centris*. These bees come equipped with specialized squeegees on their front and middle legs and coarse, oil-holding hairs on their back legs. The floral oils are carried back to the nest, where they are mixed with pollen and nectar gathered from other flowers to feed the ravenous larvae (the oil flowers don't produce any nectar themselves). In deserts as well as tropical forests, many flowers reward their bees with these nutritious secretions.

Some plants have evolved as subcontractors for bees in the nest-building business. The tropical genus *Dalechampia* has pink or greenish clamshell-like blossoms that exude resins, complex mixtures of terpenoid chemicals, which are sought out by leafcutter and orchid bees for use in nest making. In the tropical rainforests of Panama, the genus *Clusia* offers up sticky resins to stingless bees not as food but as cement for the large colonies they construct in hollow trees.

Good Vibrations: A Different Kind of Pollination

Silverleaf, or deadly nightshade (*Solanum elaeagnifolium*), grows in abundance near my home. It's a native plant hated by cattlemen here in Arizona because its poisonous foliage is reputed to kill steer or at least leave a bad taste in their mouths. But as far as I'm concerned, silverleaf is an old friend. I've watched it beckon matinal bees to its fragrant purple and yellow blossoms countless times over the last three decades. I generally get up at or before sunrise to secure a good seat to witness the drama as it unfolds. Before long, a magnificent *Caupolicana* bee, fuzzy brown with a showy black and white striped abdomen, arrives on the scene and begins methodically working its way through the silverleaf patch. Beautiful but doomed, the flowers will live for only a single day, opening at dawn and dying by sunset. Hidden within its petals are ten plump, Day-Glo® yellow anthers, each with two tiny pores on its tip. As I watch, the bee lands on a flower near me and instantly bites down on the clump of anthers. Curling its abdomen into a C, it works its bee magic, wings held closed over its thorax as it shivers its powerful flight muscles, creating sonic vibrations that are instantly transmitted into the anthers, where tens of thousands of pollen grains begin a bizarre dance. Like billiard balls gone amok, they slam into one another and the anther walls. Soon they reach escape velocity and a cloud of pollen shoots out of the pores in each anther. Some of the pollen is lost to the wind, but most sticks to the legs and fuzzy abdomen of the bee. Later, when the bee grooms herself, the pollen is packed securely on her hind legs for transport back to the hive. Before returning home, however, she will undoubtedly visit another silverleaf, whose stigma will go immediately into action, gouging out some of the stored grains. Though most of the hundreds of thousands of grains will be lost or eaten by the bees and their hungry brood, a few will land on the stigma, the first link in the chain of events that eventually yields ripe fruit and hardy seeds.

Sonification, or buzz pollination, occurs in about 8 percent of the world's 250,000 species of flowering plants. Common crops requiring sonification include blueberries, cranberries, eggplants, chilies, and kiwi fruit. Tomatoes and some legumes are especially fond of getting these good vibrations as part of the pollination process.

A Lasting Relationship: The Orchid and Its Favorite Bee

Several genera of New World tropical orchids have co-evolved with their chief pollinators, the male orchid bee (*Euglossa imperialis*), in special ways. Instead of producing nectar, these orchids secrete essential oils to lure their pollinators to pay a call. The oils are highly fragrant, with spicy scents reminiscent of cinnamon, vanilla beans, and crushed Eucalyptus leaves. When breezes waft the scent their way, the metallic green orchid bees follow it through the tangle of forest lianas to reach the perfumed sirens, usually perched high in the canopy of a leafy giant. Though they're members of the honey bee family, orchid bees are completely unlike other honey bees in both appearance and the fact that foragers can be either male or female. However, it is only the males who visit the orchids to collect the perfumed oils. Once a bee has located his potential partner, he hovers nearby, gradually edging closer. Often he is joined by other metallic green aerialists. The bees eventually land on the flower's labellum, quickly locating the glands that secrete the oils that perfume the air. Once they've landed on their prize, they are so engaged in scent collection that they can be approached by a curious bee biologist (me, for example) and plucked away with the thumb and forefinger. Like all male bees, they have no stingers. Using the foretarsi, or "bee toes," on their front legs, they busily scrape away the thin film of oil that has accumulated on the glands. These bees have developed highly specialized appendages for all possible contingencies, not unlike flying Swiss army knives.

As soon as the bees have scraped up the orchid's odorant molecules, they back off and hover a few inches from the blossom. In midair they rapidly transfer the molecules from their front legs to their middle legs and finally to the swollen tibiae of their back legs, where the molecules are sucked into spongy reservoirs and stored for eventual use.

In nearly all orchids, pollen grains are "shrink-wrapped" in waxy masses called pollinia, waiting for a pollinator of just the right size and behavior to come along and free them from their prison. The orchids have evolved an ingenious strategy to make sure this happens. While gathering up the perfumed oils, an orchid bee will sometimes lose its footing and slip on the long, narrow orchid blossoms, which function as slides that cause the bee to come into contact with the orchid's pollinia. The bright yellow pollen sacs then are neatly glued to some part of the bee's body. When the bee go-between visits another orchid of the same species in the female phase, the pollinia and their pollen are transferred—and another generation is ensured.

In my opinion, orchid bees are living jewels. My first glimpse of them was in Costa Rica in 1972. I was a college sophomore taking a course in tropical ecology sponsored by the Organization for Tropical Studies, a consortium of U.S. universities and the University of Costa Rica in San José. It was Dr. Gene Jones, a visiting lecturer in the course, who opened my eyes to the tropics and left an indelible mark on my life as a biologist.

We were in a dry, deciduous forest in northeastern Guanacaste province, at the Rancho Palo Verde, once a haven for bees and flowering trees but now a vast cattle pasture with introduced grasses prone to wildfires. I remember hiking to a secluded spot along the Rio Sarapiquí with a bottle of Eucalyptus oil in my backpack. When I arrived, I tacked a small piece of blotting paper to a large tree at the edge of a clearing and poured a teaspoonful of the clear oil onto the paper. The scent of Eucalyptus—somewhat like that of Vicks VapoRub—bit my nose with its sharp pungency. In less than five minutes, a gleaming green orchid bee, rivaling any Colombian emerald, flew into view. It was somewhat hesitant at first but eventually landed just below the oily blotter, then climbed aboard. The green bee, a male *Euglossa imperialis,* was soon hard at work scraping at the blotter with his front legs. Male orchid bees are well adapted to absorb the oils given off by the orchids. Exactly what they do with the chemicals is still a mystery. Many scientists have tried to solve the puzzle, but so far no consensus has been reached. Some think the male bees transform the chemicals into their own brand of aphrodisiac, puffing out plumes of alluring scent when they display to females. It will take a lot more sleuthing before we discover the truth behind the sex life of the orchid bee and the mystery of the purloined fragrances.

Floral Deception: Nailed by a Hammer Orchid

Orchids are sometimes a deceitful lot. In fact, many orchids are all advertisement, with nothing to back their claims. Instead of producing nectar to reward their pollinators, they don't stock their pantries with any treats at all, which means unsuspecting visitors often go home without their supper.

On the dry plains of western Australia, hammer orchids (genus *Drakaea*) send up their flowering stalks to attract the attention of passersby. This orchid has taken the art of deception to dizzying new heights, for its labellum has actually been modified to resemble a wingless female Thynnid wasp. The faux female, plump and tantalizing, sits on the end of a hinged floral arm and bobs seductively in the breeze as if it were alive and ready for an amorous tryst. The labellum has all the charms of its real-life prototype: a small shiny head, a rounded and slightly hairy body, an upturned abdomen, and the right coloring. It even produces the same blend of chemical pheromones that the real Thynnid wasp uses to perfume herself to attract a mate.

The floral charade is very effective. Male Thynnids from far and wide travel upwind, tracking the scent with receptors on their sensitive antennae. When the aroused male spies what he thinks is a willing female, he pounces rather ungallantly on her back and grabs hold with all six of his legs. But when he attempts to fly her off to a secret love nest and mate, he finds that she is unbudging. Has she capriciously changed her mind and set her heart on a handsomer wasp? While he's trying to figure this out, the orchid's plan has already been put into action. The impact made by the male as he pounces on what he thinks is a female has catapulted them both toward the orchid's floral column, where the stigma and pollinia, the orchid's sexual parts, are waiting. If the orchid is in its male phase, a pair of bright yellow pollinia likely wind up stuck to the wasp's back. Again and again the wasp tries to abscond with the "female," all to no avail. Eventually, he gets the message and buzzes off in search of the real thing. If, however, he makes the same mistake again, this time with an orchid in the female phase, his movements will throw him against the stigmatic column and the pollen grains in the pollinia will find their mark. The sexual act will have been successfully completed—though not the one the male Thynnid had his heart set on.

There are four species of hammer orchid in western Australia, all adapted to have pseudosex with four different kinds of Thynnid wasp. The relationships work because the male wasps emerge from their underground natal cells several weeks before the females. Thus, there are a lot of frustrated males flying around looking for love—prime candidates for the floral deceit the orchids practice so well.

Abuzz with Activity

Oh, to Be a Bee!

Honey bees live in a world vastly different from ours—a world I've often longed to experience for myself. What would it be like to buzz through a fresh spring morning and look down on fields of colorful wildflowers through many-faceted compound eyes? As a bee, I would actually have five compound eyes, each with thousands of slender hairs growing from its surface. Through these hairy eyeballs, I would see vibrant colors and rapid movements that are inaccessible to the sensory powers of mere human beings.

While life on the wing could be magical, it would not be without risk. My fantasy transformation could easily turn into a nightmare if I unwittingly flew into the snare of an orb-weaver spider's nest or, ironically, found myself trapped in the folds of an entomologist's net, condemned to the bitter-almond vapors of the killing tube.

In my reincarnation as a honey bee, I would savor tastes—ranging from sweet to sour and back again—not only through cells in my mouth but also through sensitive hairs on my feet. I would have no need of Kleenex, for I would smell not through a big, protuberant nose with a tendency to get red and stuffed up but through thousands of minute sensory pits scattered across the surface of my antennae. Olfaction is extremely important for us bees. Without it we wouldn't be able to locate flowers as easily, recognize our nest mates, smell an intruder in the colony, detect alarm pheromones given off by colleagues who have spotted danger, or appreciate the royal allure of the queen's substance. Floral scents cling to our hairy bodies as insistently as the scent of a woman's perfume clings to a man's jacket after an evening of dancing close. When we return to the hive after a successful shopping expedition, the clinging fragrances inform our sister workers of the kinds of blossoms we have visited during the day, a sort of travel log recording our forays out into the world. More importantly, these scents help direct new recruits to the best sources of nectar and pollen—they just sniff a returning forager with their antennae, then go out and sniff their way to the flowers.

The scents of the outside world don't mask the unique scent of the home hive, which is always there to serve as an olfactory identity badge. Guard bees sniff the body odor of workers returning from the field; if they smell something unfamiliar, the stranger is kept out or even attacked and killed if it persists in trying to enter. So if you ever wake up as a honey bee, be sure not to use chemical deodorants to mask your natural scent—it could be lethal.

Home Sweet Home: Life in the Hive

We are now going to take an intimate look at the home life of honey bees as they go about their daily routines—bringing back the raw materials they need to survive, making honeycombs, making honey, and making babies. We'll take a tour of the hive—hearth, home, nursery, pantry, day care center, and honey factory—to meet the larvae, watch the queen and her subjects hard at work, study honey manufacturing and the mating process, and find out who does what, when, and how.

If we could shrink ourselves down to bee size and enter the inner world of the nest, we would find it an alien place, dark, crowded, and oppressively hot and humid. But bees are not humans, and presumably they feel comfortable in the hive, which is home to a queen, tens of thousands of her daughters, and a few hundred or so of her sons. Double-sided hexagonal combs line the walls, floors, and ceilings of the nest. Like large corporations that house thousands of employees in diminutive cubicles, the hive is a highly regimented and efficient workplace. The waxen cubicles serve a multitude of functions, from storage pantries and nurseries to dance floors for the waggle dancing of successful foragers.

Waggle dancing, by the way, sometimes called the "ballet of bees," is one of nature's most fascinating means of communication. When a successful forager returns from the field with a stomach full of honey or bulging pollen baskets, she performs a lively dance on the combs—a little wag-tail movement followed by a straight run—that seems to be imparting information to the bees watching the show. (The spectators often beg for a sip of nectar from the obliging dancer, who takes a break to pass out free samples.)

Karl von Frisch, an Austrian animal behaviorist who won the Nobel Prize in physiology and medicine in 1973, spent decades studying what he called the "dance language of the honey bees." Von Frisch determined that the straight-run portion of the waggle dance indicated the direction of the floral treasure trove, while the vigor of the waggling, accompanied by agitated buzzing sounds, imparted the relative distance to the food. Dancing straight up on the combs indicated that the bees in the audience should fly toward the sun to find the floral patch. If the food was located in the opposite direction of the sun, the performer danced straight down.

Beeswax, the building material of the hive's interior, is biochemically expensive to make since it takes almost twenty pounds of honey, digested and transformed in the bees' bodies, to produce just one pound of wax. (Which is why beekeepers often supply man-made beeswax foundations to their colonies—they'd rather have this honey for their own use.) The wax is secreted by special glands under the abdomen of the bee, then chewed and softened until it is pliable enough to be shaped into the thin-walled hexagonal cells of the combs.

Bees on a honeycomb.

How the geometric combs are constructed by the workers without a supervisor, blueprints, rulers, or protractors remains a mystery. We believe that the precision of the hexagonal cells is the result of an instinctive behavior, which stimulates the bees to add, remove, and fashion the wax in just the right way to ensure that the walls and angles are all perfectly aligned.

The Queen and Her Consorts

The habits, the passions, that we regard as inherent in the bee, will all be lacking in her. She will not crave for air, or the light of the sun; she will die without even once having tasted a flower. Her existence will pass in the shadow, in the midst of a restless throng; her sole occupation the indefatigable search for cradles that she must fill. On the other hand she alone will know the disquiet of love. Not even twice, it may be, in her life shall she look on the light—for the departure of the swarm is by no means inevitable; on one occasion only, perhaps, will she make use of her wings, but then it will be to fly to her lover.

Maurice Maeterlinck,
The Life of the Bee

In ancient times, it was believed that honey bees came into being by spontaneous generation and that sexual acts had nothing to do with it. A passage in the King James Bible describes a honey bee swarm issuing from the carcass of a dead lion as if from a womb. From a biological perspective, this is clearly an impossibility. Most likely, the authors of the passage were actually chronicling the emergence of adult insects, probably flies, from a rotting carcass where their eggs had been laid so their larvae could feed on the decomposing tissues.

This misunderstanding of bee sexuality was enthusiastically embraced by the Catholic Church, whose faithful believe in the virgin birth of Christ. Honey bees became a symbol of this virgin birth since it was thought that they, like Christ, were the result of immaculate conception. This made the bees very special in the eyes and hearts of the devout. It wasn't until 1637 that a Dutch anatomist named Jan Swammerdam dissected queen bees and found eggs in their bodies, putting the idea of virgin bee births to rest once and for all. Swammerdam proved that honey bees were sexual creatures like other animals and needed to have intercourse if they wanted to reproduce.

In fact, not only are honey bees not virginal, they like to have fast sex on the wing, high in the air at about 15 mph. Every day, usually in the afternoon, the robust male bees leave home, ready to party. Hundreds, even thousands of drones from many colonies congregate at special aerial mating areas several kilometers from their nests, where they compete for the attention of a few dozen or so queens in order to perform their procreative duty.

The drones follow a seductive chemical trail released by the queens, who are anxious to mate. Those that overtake a queen climb onto her back, hold on tightly with their legs, and rapidly mate with her. A healthy queen mates with seven to ten drones on a given mating flight (so much for immaculate conception). She may make a number of mating flights over the course of several subsequent days, settling down to egg laying on the second to fourth day after her sexual encounters.

Once they have mated, the males fall off the queen and onto the ground, paralyzed and dying. This life in the fast lane is the destiny of many male honey bees: brief sorties chasing queens through the air, a few seconds of fun, then a rapid death. The fate of those drones who don't manage to mate is not much better. They remain in the colony, celibate, unproductive, and forced to beg for food from their industrious sisters. At the approach of winter, the poor creatures are unceremoniously evicted from the nest and left to die out in the cold.

After her mating flights, the queen returns to the colony, which she will never leave again, to begin her new role as an egg-laying machine. A longer version of her daughters, with an elongated abdomen, she can lay up to fifteen hundred eggs a day throughout her life of one to three years, for a possible grand total of six hundred thousand progeny. During her multiple couplings, she has stored several million spermatozoa in a circular, muscular organ called the spermatheca. The sperm are kept alive in the queen's body, providing a lifetime supply. Unlike Henry VIII, who needed three wives and a new religion to produce a single son, the queen bee can actively determine the gender of her offspring. When she releases spermatozoa, the fertilized eggs become sterile female workers, while unfertilized eggs become male drones with no fathers, only a mother. The eggs are fertilized in the queen's body as they travel down her oviducts and pass by the opening to her spermatheca. By controlling the muscular action of the spermatheca, she is able to regulate which eggs are fertilized and which aren't. No one knows how the queen keeps the right proportion of males to females, but she does. Generally, there are a few hundred drones in a colony and thirty thousand or more female workers—a division of gender and role that is remarkably consistent from nest to nest.

When she's ready to lay the eggs, the queen walks over the surface of the combs in the brood nest, the deepest part of the hive, looking for empty cells that have been cleaned and polished. When she finds an acceptable cell, she backs into it and lays a single, brilliant white egg, then backs out and moves on—a task that is repeated up to fifteen hundred times a day, every day of her life.

Not surprisingly, the queen receives preferential treatment in the colony, the very survival of which depends on her devotion to duty. A retinue of her daughters makes sure she is well fed and well groomed, as befits her station. Despite all the attention, however, the life of this "royal" is a lot more work than play.

The Making of a Worker Bee: From Egg to Adult in Twenty-one Days

Once the queen has deposited a fertilized egg in its cell, the female larva forms rapidly and hatches in only three days. As soon as the larva emerges from the egg, it starts feeding on a milky white secretion, called royal jelly, provided by young female bees known as nurses. The larvae are insatiable eating machines; their only imperative is to grow as fast as they possibly can. Whitish, wormlike creatures, they look nothing like the bees they will become. The head is very small, and they have no eyes, antennae, wings, or stingers. They do, however, have jaws, mandibles, and a tiny mouth through which all that food passes.

The hexagonal brood cells are left open during most of the larval development. This allows the nurse bees to make thousands of cell inspections and deposit more food as needed. The future female workers, like the future male drones, receive royal jelly for three days and are then fed a mixture of pollen and nectar known as bee bread. Larvae from fertilized eggs destined to become queens are fed exclusively on royal jelly during their entire developmental period.

On the ninth day of their lives, when they have finished eating their bee bread, the worker larvae become rigid and sit with their heads facing out of the cell opening. From tiny, paired openings near their mouths, they spin silk cocoons, getting ready for a short sleep called the pupal stage.

From day eleven to day twenty, the tissues of the larva develop into the flesh of the adult bee. Its eyes turn pink, then chocolate brown, and finally black. Like a monarch butterfly inside its chrysalis, the larva's body is undergoing an amazing series of changes.

Finally, on the twenty-first day, a new female worker bee emerges, chewing her way through the cell cap. Male drones, after undergoing a similar development process, emerge on the twenty-fourth day, while queens emerge after only sixteen days. The female "newbies" are paler than their older siblings and look like wet rats. Over the course of several hours, their hairs dry and fluff out and their skin hardens and darkens, making them look more like mature adults. Soon they figure out how to walk, using the versatile "tripod" stance of all insects—three legs off the ground and three on.

As she grows up, the female worker bee will have many different job descriptions. She will start out as a nurse bee, charged with feeding the larvae and cleaning the cells. Then, as she matures, she will be promoted to honey maker and will be assigned other chores around the hive such as making new wax cells and pampering the queen. Finally she will take flight, emerging from the nest to forage for nectar and pollen far from the safety of home. If she is born during the spring or summer months, chances are she will live for only four to six weeks—a brief, exhausting, but highly productive life.

Unlike his busy sisters, the drone has only one job—to mate with the queen and donate his genes to her stored sperm and the eventual production of new females. Once his task has been completed, his life's work is done and, without further ado, he simply dies.

"Queen Substance": The Foundation of Royal Power

Pheromones are sex-attractant chemicals given off by many insect species. In the case of honey bees, these chemicals, produced solely by the queens, are called queen substance. At county fairs across the United States, savvy beekeepers tie caged queens under their chins to attract males into a seething "bee beard," an alarming spectacle that never fails to impress the crowd.

Not only is queen substance a powerful love potion, it also helps maintain the social structure of the colony. Thanks to its allure, worker bees flock to their mother to tend to her every need. Young, vigorous queens produce great quantities of queen substance, but the level gradually wanes as the queen ages. This decrease in production may spark a supersedure, or coup d'état, by the worker bees, who replace the old queen with a younger one.

When a queen dies or is removed by the beekeeper to start a new colony, the disappearance of queen substance from the nest stimulates the workers to produce enlarged cells, stocked with plenty of royal jelly, in order to rear several new aspirants to the throne. If the aspirants emerge at the same time, they fight to the death and the survivor becomes the new queen. If one emerges before the others, she eliminates her rivals by stinging them to death while they are still in their cells. All very Shakespearean.

Foraging for Nectar and Pollen

About 25 percent of a colony's worker population consists of mature females over the age of two or three weeks, who are responsible for gathering the raw materials upon which the colony depends for its survival. Honey bees have been known to fly as far as 8.7 miles from their nests in search of pollen and nectar. Most often, however, they fly between one and three miles from home. If you draw a circle with a radius of several miles around a honey bee nest, you begin to appreciate just how much territory the colony requires in order to stock its pantries. It has been estimated that forager bees cover an area of 124 square miles in temperate regions and 186 in tropical forests.

Leaving the safety of the nest can be risky business. Wind currents and rainstorms can blow the little bees off course. Birds, lizards, robber flies, and spiders all lie in wait, ready to snare the hapless foragers in order to satisfy their own food cravings.

Most foragers venture out of the nest four to eight times a day. On each trip, the individual forager usually remains faithful to one kind of flowering plant. This constancy is good for the flowers because it means the bees transfer the pollen of one species only to other members of the same species. If bees routinely delivered the pollen of one species to the flowers of a different species, they would not be

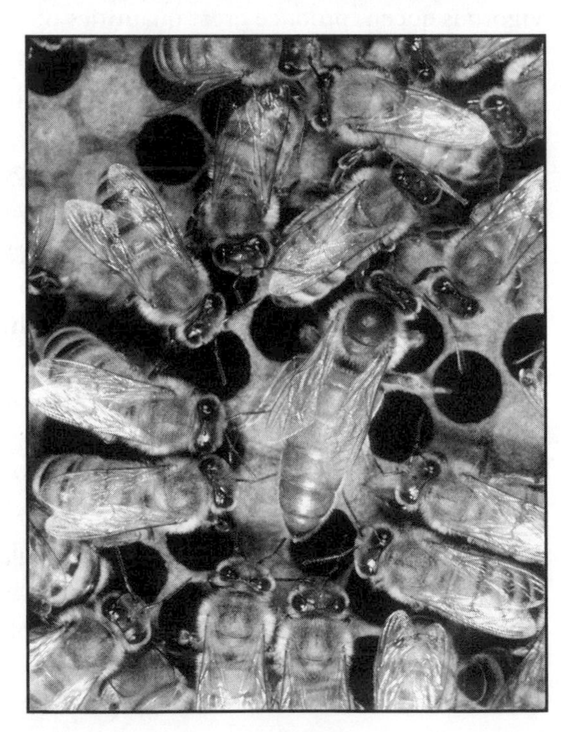

Queen honey bee and her retinue of worker bees, her sterile daughters.

very effective pollinators, and the gardens of the world would soon shrivel.

Honey bees collect nectar in a special cavity called the honey stomach or crop. To get the job done efficiently, each worker comes equipped with a strawlike appendage called a glossa at the tip of her mouth, which enables her to sip up every drop of sweet liquid from deep within the folds of a blossom.

The amount of hard work and flight time that goes into making the honey sold in a typical sixteen-ounce jar is staggering to contemplate. The contents of that ordinary container represent the efforts of tens of thousands of bees flying a total of 112,000 miles to forage nectar from about 4.5 million flowers.

How Bees Make Honey

For centuries, it was assumed that honey bees simply visited flowers and collected the honey ready-made, bringing it back to the hive and storing it there.

The truth of the matter is that honey making is an elaborate and complicated process. The first step is the collection of floral nectar from the gullets of colorful and fragrant blossoms. Floral nectar starts out as sugar water enriched with a few amino acids, proteins, lipids, phenolics, and other chemicals. While it sits in floral pools, waiting to be sampled by pollinators, the nectar takes on the aroma of the flowers that produced it. Though the scent of the nectar itself is faint, the aromas are intensified once it is concentrated into honey. Excess water is driven off and the complex volatile oils and other chemicals from the flower are magnified, becoming part of the honey and adding to its appeal. Single-source honeys reveal their characteristic aromas best at room temperature, especially when drizzled across a warm piece of toast.

When a forager returns to the hive, she looks for a nest mate to whom she can pass along her precious few microliters of nectar. The younger, stay-at-home honey makers and nurse bees often tap the returning foragers with their antennae, asking for a free taste.

The sooner the forager can unload her nectar, the sooner she can be back out in the field foraging for more. The length of time it takes the returning forager to find a willing receiver indicates the status of the honey stores in the colony. If receiver bees are eager to take the nectar, it means the colony is low on honey. If it takes the forager several minutes to find a bee to accept her offering, there is probably plenty of honey already in the combs. Or perhaps the receivers find the quality of the nectar inferior because the sugar level is too low.

When a receiver bee has relieved a forager of her nectar, she transfers it to a waiting cell. Then she stands over the cell and concentrates the nectar, evaporating the excess water by fanning her wings over the liquid and by sucking it back into her mouth and crop, then regurgitating it as many as two hundred times. In this way, she raises the sugar level of the nectar from 30–40 percent to a whopping 80 percent. The higher sugar concentration helps preserve the honey that she produces, killing microbes by sucking the moisture out of them. If the sugars fall below 80 percent, the honey will lose its ability to withstand degradation by the sugar-loving bacteria. In the United States, honey must be 80 percent sugar in order to be labeled "natural."

As soon as most of the water has been evaporated, the bees fan their wings up to 26,400 times over those sections of the comb where the concentrated nectar is stored. This fanning, along with the high temperatures in the nest, helps to thicken and ripen the nectar into honey.

Worker bees constantly check the viscosity of the ripening honey to make sure it achieves just the right consistency. When the honey is fully ripened, other workers cap the storage cells with fresh wax, ensuring that the colony will be well fed over the coming weeks and months.

Or will it? Greedy creatures beyond the hive, coveting what the bees have worked so hard to produce, frequently break into the nest to rob it of its treasure—bears, badgers, and honey hunters in the wild, beekeepers and mice when the bees live in managed colonies. Some of the plundered booty will make the long journey from honeycomb to honey jar to breakfast table, assisted by packagers, transporters, and marketers, and finally by sweet-toothed shoppers negotiating crowded supermarket aisles.

Chapter 6

Bees and Honey in Myth, Legend, and Ancient Warfare

The ancients believed that a flower's nectar came directly from heaven. Drops of firmament fell into waiting cups and bowls made of petals or were placed there by deities like Iris, the divine messenger and rainbow goddess. All the bees had to do was gather up these heavenly drops and "ripen" them in wax combs as honey. That's why Roman naturalists like Pliny called honey "the saliva of the stars."

—Peter Bernhardt, *Natural Affairs: A Botanist Looks at the Attachments Between Plants and People*

Bees and honey played a special role among the ancients. They figured prominently in their mythologies, religious beliefs, and rituals of both love and death. In fact, ancient cultures revered honey and the honey bee as magical, powerful, and even divine. As potent icons and

symbols, they provided important guidelines for both human thought and behavior.

In the Beginning . . .

"Mommy, where do babies come from?" For untold generations, children have been asking their parents this question. And adults ask themselves the much larger and more difficult question: "Where do we as a people come from?"

Curiosity about our origins is an essential part of what makes us human. The urge to discover our beginnings, to trace the roots of the tribal family tree back to the original seed, is fundamental to our sense of self. To understand our history, full of wonder and horrors as it is, we need to imagine the first act that set all subsequent acts in motion: the mysterious, time-shrouded act of creation. And once we have imagined it, we need to turn it into a story, one that can be told around campfires on windswept plains or declaimed from the pulpits of towering cathedrals or written out in sacred texts. For only in the telling and retelling will the story become real and its intrinsic lessons take on meaning.

The birds and the bees are often part of a nervous parent's answer to a child's innocent question. Stories about honey and bees are often part of the answer to the larger question about the origins of our ethnic group.

The San people of the harsh Kalahari Desert in southern Africa live in hunter-gatherer clans that range freely across their arid homeland. When they have exhausted the food reserves in a given area, the San simply pack up their few belongings and move on, if not to greener pastures, then at least to less depleted ones. To help them cope with the hard realities of their nomadic life, they have evolved a rich imaginary world, one they sometimes call the "dream that is dreaming us." Part of that vivid dreamscape is the story of how they first came into being, and a key player in that story is our friend the honey bee, a creature of great wisdom, according to the San, and one that is highly revered.

A long, long time ago, Mantis asked Bee to carry him across the dark, turbulent waters of a flood-swollen river. Bee, known for his wisdom and reliability, agreed and told Mantis to climb on to his back. Buffeted by fierce, cold winds, Bee soon grew weary and searched for solid ground on which to deposit his burden. But the stormy waters seemed to stretch all the way to the farthest horizon. Exhausted and weighed down by the much larger Mantis, Bee sank closer and closer to the lapping waves. But just as he was about to go under, he spied a great white flower, half open and floating on the water, awaiting the sun's first warming rays. Marshaling his remaining strength, Bee struggled toward the flower, laid Mantis down in its very heart, and planted within Mantis the seed of the first human being. Then, his task complete, poor Bee died. Later, when the sun had risen in the sky and warmed the white flower, Mantis awoke, and as he did so, the first San was born from the seed implanted by Bee.

As a pollination ecologist, I particularly appreciate this lovely African creation myth, with its charming story of the helpful honey bee, its mantis hitchhiker, and the brilliant white blossom that became the cradle of humanity, for the tale faithfully re-creates the actual process of pollination—bees transporting the seeds of life—on which so much of our planet's flora depends.

Far from the relentless heat of the Kalahari Desert, the ancient Romanians also cast the honey bee in a pivotal role in their creation myth, this time as the messenger of God.

God made heaven, and then, after measuring the space beneath heaven, began to form the earth. When Mole asked if he could help, God told him he could hold the ball of thread from which he was weaving the patterns of the earth. Unfortunately, Mole let out too much thread and the earth grew too large for the space under heaven. Mole was so upset by his blunder that he ran away and hid. Anxious to resolve the problem of the oversized earth, God sent his trusted messenger, Bee, to find Mole, for he wanted Mole's advice on how to rectify the mistake.

When Bee found Mole and explained his mission, Mole just laughed at the idea of a creature as lowly and error-prone as he advising the great

Creator. The clever Bee, however, hid in a nearby flower and overheard Mole mumbling to himself about what he would do if he were God.

"I would squeeze the earth," Mole said, "to create mountains and valleys that would make its surface smaller."

When Bee overheard this, he went directly to God, who followed Mole's inadvertent advice. As a result, the earth, with all its high mountains and deep valleys, now fits perfectly in its place under heaven.

Once again, Bee saved the day and the fledgling planet, for while Mole came up with the solution, forward-thinking Bee took action. Mole may be smart but, living in a hole in the ground, is no visionary. Bee, on the other hand, knows what it takes to create a world, having created the superbly functioning world of the hive.

Honey has been an essential part of the life and culture of India since prehistoric times, so it comes as no surprise that the honey bee figures importantly in the Hindu creation myth. The great god Vishnu, who in the form of Surya, the sun, created the world, is depicted as a honey bee—creator of its own ingenious world—resting on a lotus flower, the symbol of life. To help populate the newly created world, the love god, Manmadha, was armed with a bow whose string was made of honey bees, the symbol of fertility. The arrows he shot with this powerful bow spread love far and wide, resulting in lots of new babies to inhabit the land. They also inflicted sweet pain, for while the honey produced by bees represents the sweetness of love, the bee's sting represents its pain—and perhaps the pain of childbirth.

According to French anthropologist Claude Lévi-Strauss, the creation myth of the Caduveo people of the Brazilian rainforest includes a moral lesson involving honey. When the caracara (a falconlike bird who was in charge of creation) saw honey forming in huge gourds on the ground, where it was to be had for the taking, he told Go-noeno-hodi, the demiurge, to put it in the middle of a tall tree instead, so that humans would be forced to climb up and dig it out. Otherwise, he explained, the lazy humans would never do any work. So the bee might be seen as an instrument of our moral education—

a role it in fact plays in the many paeans of praise to its industry, community spirit, and chastity that appear in other cultures.

Some of the most interesting creation myths are those about the creation of bees themselves. Like many ancient peoples, the Egyptians believed that bees and honey could claim divine provenance, as this passage from the Salt Papyrus makes clear.

> *The god Re wept and his tears fell to the ground and were turned into bees. The bees began to build and were active on all flowers of every kind belonging to the vegetable kingdom. Thus wax came into being and thus was created honey.*

One can't help wondering what it was that caused the god to weep such bittersweet tears with such sweet results.

The most enduring bee creation myth is that of their spontaneous generation. Up to the seventeenth century, it was an article of faith in apicultural treatises that bees were sexless creatures, unable to reproduce naturally. They just came into bee-ing as if by magic. Out of this myth grew a tenacious sexual taboo forbidding beekeepers from engaging in carnal relations before collecting honey or trying to recover an escaped swarm. Such beliefs persist to this day in some rural beekeeping families in the United States.

The myth of autogenesis, or self-creation, by bees may have begun with the biblical tale of bees arising spontaneously from the carcass of a dead lion, from Judges 14:8–9:

> *And he turned aside to see the carcass of the lion and, behold, there was a swarm of bees and honey in the carcass. And he took thereof in his hands, and went on eating, and came to his father and mother, and he gave them, and they did eat but he told them not that he had taken the honey out of the carcass of the lion.*

As an entomologist, I find it fascinating to speculate on the factual underpinnings of this biblical account and believe it may actually be one of the earliest published records of insect mimicry. The flower fly family, Syrphidae, has many members that masquerade as bees and wasps. The sight of bee lookalikes emerging from eggs previously laid in the carcass of the lion is a more credible explanation for the biblical story than autogenesis. Or perhaps a swarm of bees searching for a new home had simply paused to rest inside the mummified remains of the predator.

The Roman poet Virgil, in Book IV of the *Georgics,* his famous treatise on agriculture and beekeeping, also refers to the spontaneous generation of bees from an animal carcass:

> *But if someone's whole brood has suddenly failed, and he has no stock from which to re-create a new line, then it's time to reveal the method by which in the past the adulterated blood of dead bullocks has generated bees. . . . First, they search out a bullock, just jutting his horns out of a two-year-old's forehead. The breath from both its nostrils and its mouth is stifled despite its struggles. It is beaten to death and its flesh pounded to a pulp through the intact hide. They leave it lying like this and strew broken branches under its flanks, thyme and fresh rosemary. . . . Meanwhile the moisture, warming in the softened bone, ferments, and creatures, of a type marvelous to see, swarm together, without feet at first, but soon with whirring wings as well, and more and more try the clear air, until they burst out, like rain pouring from summer clouds, or arrows from twanging bows.*

Food Fit for Gods

The Egyptians knew all about honey when Rome was still a village of ragged shepherds. In their magnificent temples, honey was a key ingredient in many sacred rites, for it was considered the "lord of offerings," a food fine enough to satisfy the demands of even the most capricious and willful of gods. What's good enough for the gods must certainly be good enough for kings, so Egyptian royals were buried with pots of honey to help sweeten their transition to the afterlife. Thanks to its antibacterial properties, pots of unspoiled honey are still being found by modern archaeologists excavating the rich royal tombs.

The Vedas, a collection of 1,028 hymns that constitute the sacred books of Aryan India, extol the life-giving properties of honey, the food of their gods, especially when given as a gift to a newborn male:

> *I give thee this honey food so that the gods may protect thee, and that thou mayst live a hundred autumns in this world.*

It stands to reason that honey, believed by the ancient Greeks to confer immortality, was a staple in the diet of their gods. In Greek mythology, ambrosia is the food served at Olympian dinner parties, washed down with liberal quantities of nectar. Many classical scholars believe that both items on the divine menu were derived from honey.

The Greek prefix *a-* means "not," and *mbrotos* means "mortal," so *ambrosia* actually means "not for mortals." Happily, Dionysus took pity on us and decided to share the food of the immortal Olympians with mortal humankind. (The son of Zeus and Semele, Dionysus represented the intoxicating power not only of grape wine but also of mead, wine made from honey.) Returning to Olympus after a night spent gamboling with the satyrs in an earthly forest, Dionysus encountered a swarm of bees and guided them to a hollow tree, which they obligingly filled with honey. Later, human travelers through the forest discovered the honey, and a new addiction was born.

Dionysus, known as Bacchus to the Romans, was the god of human fertility as well as of wine. The Bacchanalia, celebrated in his honor, were introduced in Rome around 200 BC by the Etrurians. The festival was originally held once a year and attended only by women. Later, admission to the Bacchanalian rites was opened to men, and the festivals took place five times a month. Things were clearly getting out of hand, and the festivals were banned by senatorial decree in 186 BC. It goes without saying that during the Bacchanalia great quantities of both grape and honey wines were consumed.

Proserpina, the Roman goddess of spring, the season when bees begin to make honey from nectar, was also the queen of the underworld. The Romans made offerings of honey to discourage her from appearing on earth in the form of volcanic lava, as she was wont to do. We can only assume that the greedy Pompeians had been stingy with their honey offerings, prompting Proserpina to visit them in the guise of molten lava spewing forth from the crater of Mt. Vesuvius.

In the Land of Milk and Honey

While the decadent Romans reveled in honey wine, the more serious-minded Jews placed a higher value on milk and honey. To people living in a harsh desert climate, a lush green landscape must have fit their idea of paradise. The pastures of this rich, well-watered paradise would be dotted with contented cows grazing on succulent grass and producing fresh, wholesome milk; the meadows would be filled with wildflowers buzzing with bees as they collected nectar and pollen to transform into golden honey. It's no mystery why milk and honey became symbols for the Jews of a blessed land.

The land of milk and honey, promised by God to the Israelites when they were enslaved in Egypt, was Canaan, lying between the Mediterranean Sea and the Jordan River. Milk and honey symbolized not only the fertility of the land but also the fertility of the people who inhabited it. In the ancient cultures of Mesopotamia, the Semitic tribes believed that honey was a gift of the earth mother Astarte, goddess of fertility, maternity, and love (and, surprisingly, war). As the land flows with milk and honey, so life-giving fluids flow from the genitals of men and women.

Not only is honey a symbol of procreation, it is a metaphor for sexual allure, as celebrated in the Song of Solomon:

Your lips distill nectar, my bride;
Honey and milk are under your tongue.

The promised land of milk and honey is also the supple terrain of a woman's fruitful body.

By the time the Jews had crossed the Red Sea, traversed the scorched deserts of the Sinai, and reached the Promised Land, they were more than ready to sample the honey they had been hearing so much about. But there was a hitch: honey is produced by bees, which Jewish dietary laws consider to be "unclean." Happily, honey-loving rabbis got around this by deciding that bees did not in fact produce honey but simply transported it ready-made from flower to hive. In other words, "clean" flowers, not unclean bees, made honey. Centuries later, in the diaspora, the Jews of Eastern Europe ushered in the new year (Rosh Hashanah) with a holiday meal that included apples dipped in honey so that the year ahead would be a sweet one. And when little boys were brought to school for the first time, they were given Hebrew alphabetical tablets smeared with honey so they would acquire a taste for learning and an understanding of its sweetness.

In the Koran, Allah, not to be outdone by the Jews, describes paradise as a land watered by mighty rivers of wine, milk, and honey. It was Allah himself who told bees to nest in tall trees and sip the nectar of succulent fruits so that they might produce honey, a divine gift he generously bestowed on humankind.

We should all give thanks for the gift of honey, for without it, our own land of milk and honey would be calcium-rich but only half as enticing.

Honey and the Secrets of Mithra

Whether it's the monotheism of the Holy Land or the polytheism of the Roman Empire, honey has always had a religious resonance. It even managed to find its way into some of the Roman mystery cults, most notably Mithraism.

The cult that worshiped the god Mithra dates back to around 1400 BC. The Indo-Iranian deity is mentioned in the sacred books—the Vedas and the *Avesta*—of both the Hindus and the Persians. Romans first encountered the worship of Mithra in the eastern provinces of their empire. Its adherents deliberately shrouded their liturgy and religious practices in mystery. Worship took place either in caves or in temples, called mithraeums, constructed to look like caves. Roman legionnaires were drawn to the cult, since Mithra was associated with bravery, manliness, and loyalty to a higher order. Women, not surprisingly, were excluded from worship.

The structure of the Mithraic cult was hierarchical. Members progressed through a series of steps or grades, each of which had a special symbol and tutelary planet. Initiation rites for new recruits as well as for cult members progressing from one grade to another were complicated and highly secretive. Much of what we know about them comes from paintings on the walls of ancient mithraeums found in Italy and the Balkans. Because fire was the symbol of Mithra's destructive and regenerative powers, initiates were not allowed contact with the archenemy of fire, water. To perform the ablutions required before entering the temple precincts, initiates purified themselves with honey, which was believed to keep a person free of evil and earthly contamination. Honey was poured over their hands to cleanse them and was also used to cleanse their tongues so that they would be incapable of speaking falsehoods. A painting found in a ruined mithraeum in Bulgaria shows a lion (symbol of one of the higher grades) identified as Melichrisus, meaning "honey-anointed."

The Power of a Symbol

Bees have been important to human beings throughout history, not only for what they produce, but also for what they have come to represent. For thousands of years we have attributed symbolic importance to bees, endowing them with mystical properties and using them as potent religious, literary, and artistic icons. We extol their virtues and hold up their lives as models for the way we should live.

In many cultures, honey bees, with their ability to fly heavenward, were thought of as symbols of the human soul. In Greece, the followers of the Orphic school of philosophy imagined that the swarming of honey bees was an outward manifestation of their belief in the "swarming of souls" toward some kind of divine unity. Muslim tradition also refers to the human soul as beelike, or as a bee itself. In

Siberia, the Buriat people believed the human soul took the form of a bee or wasp and thought it could be seen issuing from the mouth of a sleeping person. Such totem bees were not to be disturbed by molesting their colonies in the wild.

To some tribes, bee totems were a way of appropriating bee power. Members of the Juanag and Bhramada clans of India believed bee totems conferred on them the ability to control honey bees, allowing for safe and relatively pain-free honey hunts.

Egyptian kings appreciated the symbolic power of bees to such an extent that they included them in their titles, referring to themselves as "he who belongs to the bee."

Bees were so important to the ancient Maya that in their language, Yucatec Maya, the same word, *cab,* means "bee," "beehive," and "honey" as well as "earth," "village," "nation," and "world." The Maya had deep feelings for their bees not only because of the wonderful honey they produced, but also because, being stingless, the bees were viewed as touchingly vulnerable. Moreover, bee society was considered a model of the ideal human society. To the Maya, life in the hive symbolized the well-ordered interactions of a harmonious family. To be successful, a beekeeper was expected to ensure that peace and harmony prevailed in his own family. If he failed to do so, the bees, highly sensitive to quarreling and discord, might stop producing.

The Maya also believed that bees had souls and that the honey they produced had divine properties. In fact, after the Spanish conquest, the Maya celebrated mass with honey mixed with water, rather than the traditional wine, to symbolize the blood of Christ.

Bee symbolism reached a pinnacle in the early Christian church. By the year AD 395, the Roman Empire had been partitioned and Christianity had become the dominant religion throughout its vast holdings. The church during that period was rife with mysticism, taboos, and insect symbolism. Certain insects were considered icons of evil. Flies conspired with the devil and were unclean, beetles were heretics, and moths represented the temptations of the flesh. Other insects, such as bees and ants, were viewed as industrious and worthy of imitation, as suggested in the biblical passage "Go to the ant, thou sluggard; consider its ways and be wise" (Proverbs 6:6).

In a sense, early Christianity orbited around the bee just as surely as the sun was believed to orbit the earth. Honey bees were thought to live in a harmonious, well-ordered society where all worked for the common good and were governed by a wise king bee. Bee society was held up as a model that human society, with its tendencies toward anarchy and lawlessness, would do well to emulate.

Honey bees also came to symbolize the Christian ideal of chastity, for, as we've seen, they were believed to arise from virgin births, not unlike the virgin birth of Christ. As an entomologist, I find this rather amusing, for, as you will recall, a queen bee may mate with several dozen males over the course of two or three days before she takes up her egg-laying routine—hardly virginal behavior. If the church had known the facts about honey bee mating rituals, it would probably have chosen the bee as a symbol of debauchery and promiscuous lust.

Honey, or rather the six-sided cells of the honeycomb in which it is stored, figures in a legend about the Hagia Sophia, the magnificent basilica built by the emperor Constantine when he transformed the provincial town of Byzantium into Constantinople, capital of his newly Christian empire. (At the time, the Hagia Sophia was the greatest church in all Christendom; now it is one of the most important mosques in the Islamic world.) In 537, when the emperor Justinian was considering plans to restore the already time-worn basilica, he attended mass there and accidentally dropped the communion wafer. Before he could retrieve it, a bee flew into the church, seized the transubstantiated bread, and flew away. The horrified monarch sent messengers to beekeepers throughout his empire, ordering them to search their hives for the purloined wafer and offering a large reward for its return. Not long afterward, a provincial beekeeper arrived at the imperial palace bearing an oddly shaped honeycomb. A closer look revealed that the honeycomb had been constructed around the missing wafer. Justinian, always on the lookout for divine symbols, rebuilt the basilica in the shape of a six-sided honeycomb cell.

As we know, honey has long been a symbol associated with affairs of the heart. The Sumerians, as far back as 5000 BC, used honeyed words to curry favor with romantic partners. Among the world's earliest love songs is this, scratched in cuneiform on a clay tablet somewhere in ancient Mesopotamia.

Bridegroom, dear to my heart,
Goodly is your beauty, honeysweet,
Lion, dear to my heart,
Goodly is your beauty.

In many of the world's religions, honey is a symbol of veracity. While the honeyed language of love may not always be truthful, honey itself can tell no lies. The Hebrew word for bee is *dbure,* from the root *dbr,* meaning "word." Bees had been ordained by God to reveal his divine word, or truth, to humankind. Honey signifies truth because it needs no treatment to transform it after it has been collected. Truth was thought to be passed on by the bees through their honey so that the elect could express it in their scholarship and poetry. This belief spread beyond the Holy Land to Europe, where it was said that bees settled on the lips of Plato, Pindar, and St. Ambrose when they were children so that, as adults, they would speak and write only what was true. (St. Ambrose, as it happens, is the patron saint of beekeeping.) Women in some ethnic groups in the Ivory Coast and Senegal still rub honey on the lips of newborns to ensure that they will never tell lies.

The ancient Egyptians also made the connection between honey and truth in language. Their god Thoth was, among other things, the lord of books and learning, scribe of the gods, and lord of holy words, thanks to his invention of both hieroglyphics and numbers. During his festival, held on the nineteenth day of the month named in his honor, his disciples greeted one another with the phrase "Sweet is the truth" and made him generous offerings of honey.

Another reference to the connection between honey and language is found in the *Atharva Veda* of early India:

O Asvins, lords of Brightness, anoint me with the honey of the bee, that I may speak forceful speech among men.

Ancient Tales of Bees and Honey

The folk tales of many cultures derive lessons from the virtues of the wise, truthful, and industrious bee as well as from the divine power of honey. These tales have helped perpetuate our belief in the special qualities of bees and the honey they produce.

A Bee-Hunting Tip from the Aborigines

The Aborigines of Australia have always had a great passion for honey and will go to great lengths to satisfy their cravings. When a bee nest is found in the wild, the sugarbag is eagerly devoured—wax, honey, pupae, dead bees, and all. The stick that has been used to pry the sugarbag from the tree is thrown into the flames of a roaring fire, a simple act that frees the spirits of the bees to return to the heavens, where they will stay until Mayra, the wind of spring, breathes new life into the flowers. Then the bees can return to earth and produce more honey, which humans will again plunder.

Before a nest can be looted, however, it must be found. This is not always easy, as bees do not believe they were created simply to satisfy the sweet tooth of greedy men and women. They produce and store honey for themselves and their offspring, and therefore their nests are well hidden in the branches and hollow trunks of great trees. The Aborigines have several methods of discovering where the nests are hidden, but perhaps the most ingenious is the one first discovered by two Numerji brothers who lived a long time ago. (This is an elaboration of the story recounted earlier.)

The Numerji brothers were bearded giants who set out on a long walkabout through the land. At one point in their travels, they came upon a flowering bloodwood tree around which many little bees were buzzing.

"This is a wonderful thing," the younger brother said. "The insects are

scooping honey out of the flowers and flying away with it. I wonder where they are taking it."

"We'll soon find out," the elder brother said. "I will show you how to find their nest and then there will be plenty of honey for both of us."

The older brother found a leaf on which a spider had woven its web. He carefully unraveled the silken threads of the web, then climbed up into the branches of the bloodwood tree.

"I am going to put bits of the sticky web on the bees," he explained to his brother. "You will be able to see them clearly now. Watch where they go."

The younger brother went off in pursuit of the tagged bees and soon came running back panting with excitement.

"I have found it," he cried. "They fly into a hollow tree down there. That's where the nest must be."

The elder brother climbed down from the bloodwood tree and together they went to the hollow tree. They broke the bark with their clubs, chopped out the sugarbag, and ate it greedily.

Thanks to the brothers' clever stratagem, the Aborigines learned the art of attaching a tiny white thread or scrap of material to the honey bees to guide them to the bees' nests.

How Elephants Got Their Trunks and Bees Their Hives: A Fable from Thailand

This story explains why elephants look the way they do—and why the best way to steal honey is to chase the bees away with smoke.

In ancient times, elephants did not have the long trunks they do today and bees did not live in nests in hollow trees. Instead, they built their nests on branches in the open air.

One year the rains were extremely meager and the land became dangerously dry. The elephants found it increasingly difficult to find enough leaves to feed on. The bees were also having trouble collecting the nectar and pollen they needed, as all the flowers were dying.

Finally, as dry as tinder, the forest caught fire. The elephants tried to outrun the danger, but the lumbering creatures soon grew tired as the flames spread unchecked. When they called for help, the bees offered to lead them to safety in return for free transport. The elephants opened their mouths, and the bees flew inside to escape the hot air and choking smoke. They settled in the elephants' short snouts and from there directed their companions to a nearby lake. The elephants waded into the middle of the lake and stayed there until the fire had spent itself.

It was now time to leave the lake and resume their hunt for food, but the bees had become accustomed to the cool, dark interior of the elephants' snouts and began building their hives there. The elephants bellowed and trumpeted in rage and began to exhale mightily in order to evict their unwanted lodgers. After several hours of trumpeting and exhaling, their snouts had stretched into full-sized trunks, but the bees remained stubbornly inside.

The elephants finally decided that since the bees had flown inside their snouts to escape the smoke, smoke would be the best way to get them out. So they walked into the still-smoldering ashes of the fire, inhaled deeply, and held the smoke in their mouths and trunks until the bees had fled. Then they returned to the lake to drink and cleanse their palates. Thanks to their new, improved appendages, they could reach the water without having to stoop.

The evicted bees, having become very comfortable building their hives in cool, dark places, searched for something similar and found that the next best thing to an elephant's trunk was the hollow trunk of a tree. This is why an elephant's nose and the body of a tree are called trunks, and why bees who live in hollow trees are called phung phrong *or "elephant's mouth bees."*

The Song of Rinn: A Scottish Folktale

In many Scottish folktales, honey is associated with desire, seduction, and death. One such tale tells the story of Rinn, a beautiful youth whose bewitching song is called the "Honey of Wild Bees."

On a visit to the world of the living, Rinn, the Lord of Shadow and Lord of the Land of the Heart's Desire, met Bobaran the Druid and the beautiful maid, Aevgrain.

"I am come here," Rinn told Bobaran, "because I follow the shadow of my dream."

As he listened to Rinn, the old Druid thought he had never heard a voice so sweet. At dusk, when Rinn sang his song of love, Bobaran saw a forest glade filled with moonlight. Soon the image lulled him to sleep. While Bobaran slept, Rinn gazed at Aevgrain, whose eyes were shining upon him as two stars.

"Play me no sweet songs, O Rinn," she murmured, "for already I love you."

Rinn smiled, but he touched the strings of his harp.

"O heart's desire, my delight!" he whispered.

"O heart's desire!" she repeated. Then her white hands moved like swans through the shadowy flood that was her hair and she leaned forward, looking into the eyes of Rinn.

"Tell me who you are, whence you are," she said.

"Will you love me if I tell this thing?"

"You are my heart's desire."

"Will you follow me if I tell this thing?"

"Yes."

"I am called Rinn, Honey of the Wild Bees. I am the Lord of Shadow. But here, O Aevgrain, my name is Death."

Honey—the sweetness of desire, the longing of love, the seductiveness of death. Despite what Rinn told her, Aevgrain followed him to the Land of Shadow, where she fell into a sweet dream from which there was no awakening.

The Honeymoon Controversy

From tales of fatal love, let's journey to a place where love leads to marriage, and marriage to a happy honeymoon.

In the far north of Europe, fierce pagan gods once ruled strapping warriors who plundered the world—and plundered bee nests as well. For as bloody-minded as the Norsemen were, they had a sweet tooth. They also had a sweet sense of the romantic. Not only did they harvest honey and turn it into wine, they conceived the idea of the very first honeymoon. Or so their descendants, the modern Scandinavians, claim.

A closer look reveals that there are other claimants to the distinction. Among them, the case of the Druidic Celts is particularly convincing. In Celtic culture, each lunar month was given a name. May was the Honey Month, the time of year when honey was collected to be fermented into honey wine or mead. Its moon was called the Honey Moon. Celtic tradition dictated that all marriages be performed on May Day. After the wedding ceremony, the newlyweds remained in seclusion throughout the period of the Honey Moon, getting to know each other and laying the foundation for years of future bliss. The flowering of love was abetted by the copious quantities of mead which the young couple consumed. Once the Honey Moon waned and the mead stopped flowing, young love had to face the exigencies of the real world—work, children, in-laws, and arguments. But steeped in honey during the honeymoon, the newlyweds had hopefully absorbed enough sweetness to withstand the bitterness that life sometimes brings.

Honey and Bees as Weapons of Limited Destruction

When a marriage goes sour, the spouses may fling stinging words at each other. That's bad, but there are worse things one might have to endure. Imagine, for example, stinging bees poured down on your head from the battlements of a fortress you are trying to take. Not a happy thought. But the fact is, stinging honey bees were often used in warfare to repel obstinate invaders. Roman armies catapulted so many bee nests on their enemies that they created a paucity of hives during the later years of the empire, which resulted in a protracted shortage of honey.

The ancients even used honey to practice a form of chemical warfare on their unsuspecting enemies, azalea honey being the secret weapon. When Mithridates VI instigated a revolt of the Black Sea kingdoms against Roman rule near the end of the first century BC, the great general Pompey was sent to quell the rebellion. Soon after his soldiers arrived in the region, they discovered that jars of azalea honey had been left for them as "tribute." The troops gorged on the honey and soon fell ill from the powerful alkaloids found in the nectar of the Pontic azaleas. Too sick to fight, the legionnaires were easily slaughtered by the crafty, beekeeping "barbarians." Eventually, Rome regained control of the Black Sea area, but only after suffering a humiliating defeat—one that may have had its seed in Greek literature.

Mithridates VI, who had received a Greek education before ascending the throne, had probably read the works of Xenophon, who, with ten thousand Greek mercenaries, had beat a hasty retreat from Persia to Asia Minor around 400 BC—some three hundred years before the Roman defeat in the same part of the world. After his return to Athens, Xenophon wrote an account of his misadventures with the mercenaries, which became required reading among Greeks of the educated class. Xenophon's *Anabasis* describes the Greeks' harrowing flight from Persia after their participation in a failed attempt by Cyrus the Younger to overthrow the king. Cyrus, a jealous sibling of the rightful occupant of the throne, had hired the Greeks to help him seize power, and he requested that Xenophon, a noted disciple of Socrates, accompany the mercenaries to record the glorious events as they unfolded. Unfortunately, the events that actually did unfold were far from glorious. Cyrus was killed in battle near Babylon, at which point the Greeks realized that they stood a good chance of being slaughtered—and weren't even going to be paid for their trouble. They chose Xenophon to lead them on the long, arduous journey back home to Greece. At one juncture during their retreat, they set up camp on the shore of the Black Sea and, while scavenging for food, came upon a store of wild honey made from the nectar of the Pontic azalea. Who knew that honey could be poisonous? But this honey was, for Xenophon wrote that those who ate it lost their senses and suffered from dreadful vomiting and purging. Four centuries later, Mithridates put that information to good use.

During the eleventh century, at the dawn of the age of heraldry, the bee's role in warfare shifted from weapon to icon as it became a symbol of valor rather than an instrument of victory. Ambitious

Toxic Honey

Dare I use the words *toxic* and *honey* in the same breath? Yes, for the Pontic azalea honey used in ancient times to debilitate attacking enemy soldiers is not the only poisonous form of honey.

While it's extremely unlikely that you'll ever encounter a toxic honey, they do exist. Sometimes, though very rarely, they are inadvertently mixed in batches of nontoxic, locally produced honey in quantities large enough to make someone sick. Even more rarely, a toxic honey might mistakenly be sold as wildflower honey at a roadside stand or farmers' market.

When bees in the southeastern United States visit the beautiful, deep yellow blossoms of the Carolina jasmine, the result is toxic honey. In the foothills of the California Sierra Nevadas, the local horse chestnut yields a honey that can sicken both bees and humans.

Other unfriendly honey plants in the United States include loco weed, southern leatherwood (not to be confused with the wonderful Tasmanian leatherwood), rhododendron, and Jamestown or jimpson (jimson) weed, which contains heart poisons and alkaloids. Tansy ragwort has also been shown to have toxic materials in its nectar. Beekeepers and local honey lovers need to be cautious if they live in areas where these plants grow in abundance.

You may wonder why a plant would poison its own nectar. But the fact is, what's toxic for some isn't toxic for all. Many of these toxic plants co-evolved with pollinators that can cope with alkaloids and other poisons. The plants are probably spiking their nectar with nasty chemicals to discourage visits from nectar-robbing, nonpollinating insects and animals.

knights devised crests for their coats of arms that displayed real and mythic beasts—brooding lions, leaping stags, fanciful unicorns—as well as our friend the honey bee, who symbolized order and obedience to a higher power.

Maffeo Barberini (1568–1644), who would later become Pope Urban VIII and play a vital role in fostering the Renaissance, replaced the three houseflies that had originally adorned his coat of arms with three bees, which is why modern-day visitors to Rome are likely to see the three bees featured in monuments, paintings, and sculptures all over the city. The bees on Barberini's upgraded shield were meant to signify his industriousness.

About two hundred years later, Napoleon I, perhaps inspired by medieval heraldry, replaced the fleur-de-lys as the symbol of France with the honey bee, invoking it as a symbol of industry and sovereignty. The image of the bee figured prominently in the Napoleonic

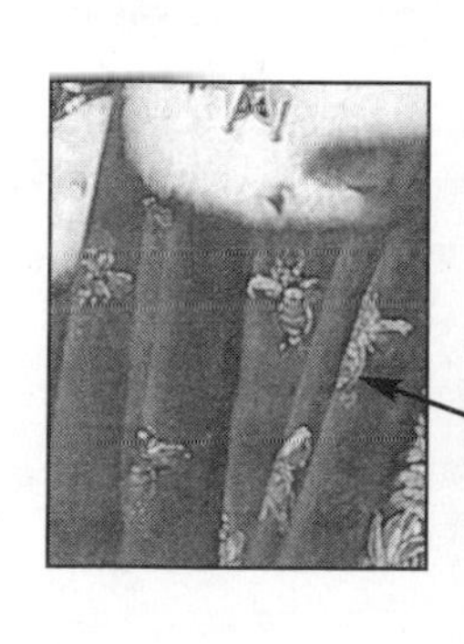

Napoleon I in coronation robes decorated with honey bee motifs. Closeup of bee symbols on Napoleon's robes.

coat of arms, reminding the revolution-prone French of their duty to settle down and become hardworking *citoyens* (or better yet, worker bees) of the new empire. Following the imperial example, many U.S. states,including Utah, Georgia, Kansas, Maine, and Missouri, have designated the honey bee as their state insect.

In our day, insect heraldry has been put to quite a different use. No fewer than thirty-five college teams have adopted mascots and symbols that have six legs. Surely you've heard of the Georgia Tech Yellow Jackets, not to mention professional teams such as the New Orleans Hornets and the Sting, Chicago's renowned soccer team.

Since victory on both the battlefield and the playing field depends on the individual's submission to the will of the group, what better example to put before the rank and file than that of the honey bee, willing to sacrifice its life for the good of the colony? Obey your general, obey your coach, and, above all, obey your queen.

Chapter 7

Trading Honey in the Ancient and Modern Worlds

Instead of dirt and poison, we have rather chosen to fill our hives with honey and wax, thus furnishing mankind with the two noblest of things, which are sweetness and light.

—Jonathan Swift, *The Battle of the Books*

The role of honey as food and medicine, as well as potent religious and political symbol, made it a highly lucrative item in world commerce. Everyone wanted honey, but not everyone could produce it—so enter the honey merchants and traders, those savvy individuals who satisfied the demand for honey by facilitating its journey from hive to market, thence to temple, apothecary, and cooking pot.

Trading Honey in Ancient Cultures

Egypt

People have been keeping bees and trading the honey they produce for thousands of years. The Egyptians, who were among the first to domesticate bees, transported

their sweet cargoes on barges that plied the Nile. The importance of the honey trade in Pharaonic Egypt was confirmed in 2002 at Saqqarah when archaeologists unearthed the opulent tomb of Ubi, the chief supervisor of beekeepers. His was an important job, for the demand for honey was great throughout the whole of the Egyptian kingdom. It was used to sweeten many foods and beverages, was fed to sacrificial animals to sweeten the pleasure of the recipient god, and was used in medicines to keep people alive and well. It was also associated with love and romance, just as it is today. In one Egyptian marriage contract, the groom pledged to give annual gifts of honey to his bride, perhaps to sweeten the course of conjugal life: *I take thee to wife . . . and promise to deliver to thee yearly twelve jars of honey.*

Scenes of beekeeping and honey trading have been found at archaeological sites throughout Egypt. Reliefs and hieroglyphs on the walls of many tombs vividly depict the established methods of managing apiaries for large-scale honey production. The earliest images, dating back to the Fifth Dynasty, around 2500 BC, were unearthed in the ruins of the Sun Temple of Ne-user-re at Abu Ghurab, near modern Cairo. The once brightly painted reliefs show workers blowing smoke into large cylindrical clay hives to calm the bees while their keepers harvested the honey. By 1184 BC, when Ramses III ascended the throne, honey was being produced in vast quantities to meet the growing need.

The amount of honey used was staggering. One record indicates that Ramses III, whose reign lasted until 1153 BC, made gifts of twenty-one thousand jars of honey to Hapi, the god of the Nile, and ordered another 7,050 jars to be used in making the honey cakes the deity was partial to. This huge amount of honey required the total annual production of at least five thousand hives. When local supplies fell short, honey was imported from Phoenicia, in modern Lebanon, and Canaan, the biblical land of milk and honey in what is now Israel.

Methods of beekeeping and honey trading seemed to change little during the millennium that followed the death of Ramses III. But after Egypt fell under the domination of Rome in the first century before Christ, a new dimension to the honey trade gradually evolved:

itinerant beekeeping. This new way of managing bees was described by Arab historians who wrote during the Middle Ages, when the practice was no longer new but still thriving. Every year at the end of October, when the wildflowers came into bloom along the banks of the Nile, itinerant beekeepers loaded their clay hives onto small papyrus rafts or large barges and floated them downriver to follow the nectar flow north (a practice similar to that of mobile beekeepers in the United States today, who load their hives on flatbed trucks to follow the flow). Since the bees could continue foraging and producing over the course of the journey, mobile beekeeping ensured the enterprising merchants the largest honey crops possible. In February, when the bloom began to wane, the floating apiaries ended their journeys dockside at Alexandria or one of the other cities of Lower Egypt. There the hives were offloaded and their honey sold. The floating bee cities, complete with honey-making factories, must have been quite something to see.

The Indus Valley

By 2000 BC, Mohenjo-Daro and Harappa, the first cities established in the Indus Valley in what is now Pakistan, were thriving centers of commerce. We know little about the culture of the Indus Valley peoples, since their language has not been deciphered and, unlike Egypt, no temples or palaces with visual records painted on the walls have survived. But we do know that trade was very important to them. From remains found in excavated tombs, including food traces, jewelry, clay jars, and fired seals probably used by merchants to close their deals, we surmise that traders dealt in gold and metalwork, pickled fruits, vegetables, wine, and honey. These items were exchanged in neighboring Mesopotamia and India for incense, pearls, and dates, among other goods. There is no evidence indicating what the Indus Valley people did with their honey beyond trading it, though it was probably used as a sweetener and may have been fermented into honey wine, or mead.

The Indus culture collapsed around 1000 BC, perhaps under the pressure of invading Indo-Aryans from Central Asia. Before long, the victorious Indo-Aryans had established themselves throughout the northern Indian subcontinent. The newcomers were Hindu and the honey trade is mentioned in the *Rig Veda,* the collection of sacred texts composed by Hindu philosophers between 1500 and 900 BC. In one such text, priests of the Brahmin caste were expressly forbidden to trade in honey, perhaps due to fears of a conflict of interest, since honey was an important element in the religious rites they conducted, a much-valued offering to the gods. As the *Rig Veda* suggests, the gods were not alone in their appreciation of honey.

> *Limitless earth, whom the gods, never sleeping,*
> *Protect forever with unflagging care,*
> *May she exude for us the well-loved honey.*
> *Atharva Veda,* 19:52

Persia

Persian civilization, dating back five thousand years, is one of the oldest and richest in the world. Persian merchants engaged in a lucrative honey trade with the conquered peoples of their vast empire, which at its peak during the reign of Darius the Great, circa 550 to 486 BC, stretched from the Nile in the west to the Indus in the east. Beekeeping had been practiced in Persia for hundreds of years before the reign of Darius, and honey had long been an important part of the Persian diet. A typical breakfast (then and now) might have included eggs, flatbread, butter, goat cheese, and honey. Among the nobility, roasted chicken or pigeon stuffed with walnuts, onions, pomegranates, and honey was often the main course on feast days honoring their god Ahura Mazda.

In an important ritual performed during Nowruz, the pre-Islamic Iranian new year celebrated on the first day of spring, people ate foods sweetened with honey to ensure good luck for the year to come. A traditional Nowruz favorite was *sohaan-e asali,* a dessert made with honey, saffron, almonds, and pistachios.

The Persians also mixed honey with water and allowed it to ferment into mead, an alcoholic drink popular in the days before Islamic prohibition. Mixed with herbs, they used honey as a medicine to treat open wounds, sore throats, and other common ailments. When members of the nobility died, their corpses were rubbed with honey and encapsulated in an outer layer of beeswax to preserve them.

Greece and Rome

As mentioned earlier, the Greeks believed it was Dionysus, the god of wine, who bestowed honey on humankind. Always enterprising, the Greeks made good use of this gift, successfully trading honey throughout the classical world during the fifth and fourth centuries BC, particularly with the Persians, Assyrians, and Phoenicians. At ports such as Piraeus, near Athens, they stored honey in barrels and loaded it onto merchant ships that crisscrossed the Mediterranean and Black Seas, exchanging the honey, along with silver, olive oil, wine, and wheat, for cinnamon, pepper, silks, and timber. Honey played such a key role in Greek commerce that it was featured on coins along with the bee, a revered symbol of industriousness and efficiency.

Throughout the course of their empire, the Romans also minted and circulated coins that depicted the industrious honey bee, enabling people to purchase their favorite sweetener with money that featured its maker. Honey was the primary sweetener during Roman times, since sugar was unknown in the Mediterranean world. It was used in cooking, in preserving meat, and, when mixed with water and wine and allowed to ferment, in a form of mead called *mulsum*. *Mulsum* was so inexpensive to make that the government doled it out to the common people to fire up their patriotism during public events. In the first century AD, *mulsum* was being produced in such quantities that the demand for honey, one of its principal ingredients, increased dramatically, a welcome occurrence for both Roman beekeepers and honey traders.

The Arab World

By the seventh century, the Arabs, who had recently emerged from the deserts of what is now Saudi Arabia to conquer and convert much of the known world, were well established as the leading traders in the Middle East, North Africa, and much of eastern Asia. Baghdad, the capital of the Muslim caliphate, was the center of a culture noted for its great works of literature and philosophy as well as important scientific and mathematical discoveries. Arab merchants led caravans across the treacherous dunes of the Sahara Desert to barter citrus fruits, almonds, pistachios, saffron, and honey for the precious cakes of salt that were sold by the tribes of the West African Sahel. Although mead was used only for medicinal purposes in the Arab world, honey was a key ingredient in many of their favorite dishes, including pastries and roasted fowl adopted from the cuisine of the conquered Persians.

China

The first evidence of organized beekeeping in China dates from the second century AD, considerably later than its establishment in the Mediterranean world. But even then, it was conducted on a relatively small scale, and most of the honey used in China for culinary and medicinal purposes was imported from the Mediterranean and the Middle East along the fabled Silk Road.

This great trading route, stretching from Byzantium in the West to China in the East, was actually a network of well-worn footpaths and rutted dirt roads connecting remote outposts and bustling markets in the far-flung regions of the ancient world. From the first century BC to the thirteenth century AD, when Marco Polo traveled it from Venice to the distant capital of the Mongol general Kublai Khan, camel and donkey caravans made their way along the Silk Road, crossing sun-scorched deserts and rugged, windswept mountains and encountering hostile tribes, rapacious bandits, and outlandish customs, all in the name of commerce.

Rice, sheep, dried fruit, tea, and wine traveled the Silk Road. Horses, glass, lacquer, cotton thread, ivory, gemstones, wool, and linen also went along for the ride. Honey was generally transported from the Mediterranean to China in dried gourds or bags made of animal skins like those used for wine. In Samarkand, an important stop on the Central Asian leg of the road, the Chinese bartered their highly prized silk for the highly prized food of the honey bee.

Not only did a myriad of goods travel the Silk Road, but culinary knowledge passed from region to region as well, introducing new flavors into old cultures as exotic spices were incorporated into local dishes. Many of these recipes called for honey, for the new, international menu featured honeyed wines, meats, and desserts.

Russia

In the eleventh century, medieval Russia became a major player in international trade. The important trade routes that networked south from Scandinavia to the Byzantine Empire and east to merge with the Silk Road all crossed the great Russian steppes. Russian rivers such as the Dnieper and the Volga served as highways ferrying Russian and Viking goods to the teeming markets of Constantinople. The Byzantine capital became the major hub of Russian trade, where Russian honey was exchanged for such luxuries as porcelain, glassware, and jewelry. Most of that honey was used as a sweetener or in medicines, but some was diluted with water, fermented, and made into mead. Beeswax, which was often traded along with honey, was used to manufacture the thousands of candles that illuminated the frescoes and icons that filled the dim interiors of the Eastern Orthodox churches.

Much of the prized Russian honey originated in the vast boreal forests of the interior and was widely traded within the country as well as abroad. Merchants from the city of Novgorod crossed the icy wilderness on skis or in horse-drawn sledges in order to collect honey from local tribes. These tribes may have hunted honey as far back as prehistoric times, a practice that by the Middle Ages had evolved into forest beekeeping. Enormous amounts of honey and beeswax were produced in these great forests and harvested by tribal peasants. Whatever honey wasn't sold abroad was used primarily for mead making.

Once the Russian honey had made its way south to the great

bazaars of Constantinople, Damascus, Baghdad, and Shiraz, it was sold along with silk and other cloth from China, pepper and spices from the East Indies, salt, herring, and beer from the Baltic, woolens from England and Flanders, and olive oil and wine from the Mediterranean.

Like the Silk Road, this trade also brought disparate cultures into contact and facilitated intellectual and artistic cross-pollination. The Russians imported Greek Orthodoxy from Byzantium, along with the silks and jewels that embellished both its clergy and the aristocracy.

Central America

Long before the Russians were exporting honey and importing a new religion, the Maya were trading honey and beeswax along the routes that united their great cities throughout southern Mexico and Central America. Honey, as we've already seen, was vital to the Maya not only as a cooking ingredient and medicine but also as an important part of many sacred rituals. Other goods traded included salt, ceramics, jade, obsidian (hard volcanic glass used in tool making), and the iridescent quetzal feathers that were prominently featured in their religious and royal paraphernalia.

The Rise of Sugar and the Decline of the Honey Trade

Sugarcane is a giant grass native to the region of the Ganges delta in India. According to at least one legend, Buddha came from the land of sugar, called Gur, in present-day Bengal. The *Ramayana,* a Sanskrit epic from the twelfth century BC, tells of glorious banquets with "sweet things, syrup, canes to chew." When the Persian conqueror Darius the Great invaded the Indus Valley in 516 BC, he found the locals growing a "reed that gives honey without the aid of bees" and brought some samples home. Slowly, over the course of the following centuries, conquests by other foreign invaders as well as the Arab caravans that plied the Silk Road introduced sugarcane throughout the Middle East. However, it remained a relatively rare commodity both there and in Europe, so expensive that only the wealthiest could afford the "Indian salt."

Honey maintained its position as the primary sweetener until the early sixteenth century, when everything began to change. In 1506, Pedro d'Arranca brought sugarcane plants to the island of Santo Domingo (today's Dominican Republic), where sugar production became a major industry and soon spread throughout the West Indies. Grown by slaves on large plantations, sugar was cheaper and easier to produce and transport in large quantities than honey. By the seventeenth century, sugar had found its way into global commerce, and the volume of honey that was traded declined drastically.

While sugar was the major culprit in the decline of honey and the honey trade, it was not the only one. Honey had been used for healing purposes throughout the ancient world for thousands of years. But well before the introduction of cheap sugar, the therapeutic applications of honey had fallen out of favor among the physicians who had displaced traditional priests and shamans in the practice of medicine. It is difficult to say for sure why this happened, but by the Renaissance and the age of scientific enlightenment in the eighteenth century, honey was considered an old-fashioned folk remedy. New medicines and medical procedures began to emerge and, though primitive by modern standards, won the trust of the scientific community.

As honey was losing its place in the medicine cabinet, it was also falling out of grace in religious ceremonies. The rise of Christianity and later Islam put an end to most of the pagan rituals in which honey had played a leading role. The old gods were replaced by a single deity who seemed to have little interest in offerings of any sort, including sweet, golden honey.

At about the same time that cheap sugar flooded the markets of Europe, supplies of honey began to dwindle. As human populations increased, so did the acreage they farmed and the cities and towns they inhabited, encroaching on traditional "bee pastures" and leaving bees with fewer and fewer foraging opportunities. Less nectar meant less honey, less honey meant higher prices, and higher prices meant an all-time low for the honey trade.

In the nineteenth century, beet sugar joined cane sugar as a cheap sweetener, and the honey trade was dealt yet another serious blow, from which it would not recover until the latter half of the twentieth century.

Trading Honey Today: The Bitter Consequences of Globalization

The Silk Road of the twenty-first century is globalization, which has made our ever-shrinking planet shrink even further. Governments of the major industrialized nations promote globalization and reward participating developing countries with subsidies and favored status as trading partners.

These liberalized trade policies have had a profound impact on the way we do business and how economies are measured. In ways that may not always be apparent, globalization affects the quality of our daily lives. The deteriorating quality and taste of many of our foods reflect recent changes in the stewardship of the earth due to the growth of modern industrial agriculture.

The honey trade is one of the many victims of a worldwide economy that has grown too large and too fast, without adequate monitoring and controls. The politics of globalization are rarely aligned with the needs of small beekeepers, small honey producers, and individual honey lovers such as you and me. Today we have far fewer choices when we go to the supermarket to buy our favorite sweetener, for the homogenizing aspect of global commerce has resulted in a less diverse product, still sweet but somehow not as flavorful and appealing as the thick, golden liquid we knew before.

Today's honey trade is still relatively small compared to that of sugar, with about 2.64 billion pounds of honey consumed annually worldwide, versus 326 billion pounds of sugar. Nonetheless, it is now truly international in scope. What begins on a warm day in early spring, when nectar is carried from flower to hive on a journey powered by the transparent wings of a tiny bee, ends when huge containers of honey are loaded into massive eighteen-wheelers and giant oceangoing freighters, destined for all parts of the world.

Large-scale exporters, including China, Argentina, Australia, Russia, and the United States, dominate the world honey market. Among these countries, China ranks first, having sold 117 million pounds in 2003 (quite a change from ancient times, when imperial China imported most of its honey). From April 2003 to April 2004, the United States imported 22 million pounds of honey from China, 9 million pounds from Canada, and 2 million pounds from Mexico, with another 26 million pounds coming from other countries. Beekeepers and their advocacy groups, the American Honey Producers Association and the American Beekeeping Federation, have lobbied against the importation of inferior, less expensive honeys for decades, so far without success.

One danger of globalization is that when something goes awry in one part of the world, its effects are felt everywhere. A good example is what happened in 2002 in the global honey trade. An antibiotic, chloramphenicol, was detected in Chinese honey exported to the United States and other countries. This antibiotic is used to control diseases in shrimp, crayfish, and honey bees. In humans, chloramphenicol is used to treat life-threatening infections when little else works. It is administered sparingly, however, since a small percentage of the population can be affected by a rare but potentially lethal side effect, idiosyncratic aplastic anemia. Canada, the United Kingdom, and the United States immediately banned all Chinese honey products, obliging merchants to pull the honey from their shelves. Though these bans were later lifted when consignments of honey tested free of chloramphenicol, the risk to consumers was a very real one.

Trouble for the Honey Bee and for Us All: The Globalization of Pesticides

The policies of international institutions such as the World Bank and the World Trade Organization as well as treaties such as the North

A Terrifying Episode in the Global Honey Trade

The deserts of the Middle East boast grand expanses of seasonal wildflowers, especially during years when the rainfall is more abundant than usual. Bees take notice of these floral bonanzas, producing a high-quality honey that is eagerly purchased by merchants and distributors throughout the region.

In 2002, investigative reporters from *Newsweek* magazine uncovered a gruesome secret about the Middle Eastern honey trade. Afghanistan has long been famous for the quality of its desert honey crop. Village beekeepers set their hives out in remote mountain canyons and on hillsides where flowers grow. Aside from opium, honey is one of the few items exported from this war-ravaged, desperately poor country. The opium trade, of course, has long been illegal—and now the honey trade had taken an illegal turn.

Starting in the 1990s, one of Osama bin Laden's top aides, Abu Zubaida, claimed to be a honey merchant and traveled frequently from Afghanistan to the city of Peshawar in Pakistan. There, while selling his honey to commercial distributors, he screened potential recruits and raised funds to support the global terrorist network. But that's not all. Forged documents, weapons, ammunition, opium, and large amounts of cash were carefully bagged, then concealed inside fifty-five-gallon barrels of honey for distribution to groups throughout the Middle East. Who would ever have suspected that our beloved honey could be inadvertently drawn into the service of international terrorism?

Thanks to the journalistic exposé, authorities soon put a stop to this subversion of the honey trade and vowed to be more vigilant in the future.

American Free Trade Agreement (NAFTA) encourage the production of cash crops for export, which in turn increases the use of dangerous pesticides to ensure large yields. DDT, for example, which is banned in the United States, is sold by American manufacturers to less stringent developing countries. (Remarkably, no country that bans certain pesticides within its own borders prohibits the export of those same chemicals.) With multinational food purveyors lobbying successfully to further the cause of global trade and an industrialized agricultural model, pesticides have become big business: worldwide sales in 1995 were over $29 billion and are still on the rise. During 1998 and 1999 (the most recent years for which data are available), the Environmental Protection Agency reported that the total amount of pesticides used in the United States was five billion pounds.

The toll these pesticides have taken on the bee population is tragic. Bee kills, with "aprons" of dead bees littering the ground in front of their hives, are occurring with increasing frequency, caused by the drift of airborne pesticide sprays, the use of microencapsulated pesticides that infiltrate plants and poison their nectar and pollen, and the resulting contamination of beeswax and honey.

Bland, tasteless honey, an ecosystem at greater risk than ever, our longtime friend and source of food and inspiration, the honey bee, in increasing danger—these are just some of the consequences of global trade, which the world's governments and its citizens would do well to take into serious consideration.

For the Best Honey, Go Local, Not Global

Unfortunately, most of us have yet to experience the rich flavor and evocative aromas of local, or "boutique," honeys purchased directly from beekeepers or at a farmers' market. As you will see in Chapter 8, there is an enormous and enticing variety of honey out there for you to discover and enjoy. In this age of globalization, however, the honey we find on our supermarket shelves has often come from halfway around the world, displacing the local and regional honeys that were once widely available. Blended with other honeys, these generic honeys have been robbed of any unique regional or floral identity. Transported in large steel drums by rail or in the bellies of freighters, they arrive at their destinations already past their prime.

But, like consumers everywhere, we can buck the trend by insisting on more food choices. Ask your store manager to stock tupelo, blueberry, orange blossom, Eucalyptus, or other unblended honeys. Or follow the example of ethnobotanist Gary Paul Nabhan, my friend and colleague. As Gary said so convincingly in his book *Coming Home to Eat: The Pleasures and Politics of Local Foods,* there are numerous nutritional, economic, and taste advantages to buying locally or regionally produced foods. This certainly applies to honeys. In our global world of homogenization, knowing where your food comes from can change your life. By searching out and buying local and regional honeys, you'll be supporting local beekeepers and small distributors as well as giving yourself an unforgettable treat.

Chapter 8

A Taste of Honey: Sampling Varieties from Around the World

Honey is the flower transmuted, its scent and beauty transformed into aroma and taste.

—Stephanie Rosenbaum, *Honey: From Flower to Table*

There are many varieties of honey for you to experience, each with its own color, aroma, and unique, distinctive flavor. This chapter is your guide to the delicious and surprising world of honey. I'll introduce you to popular types of honey in the United States, then look at some unusual honeys from other lands. And I'll tell you how you can throw a honey-tasting party yourself and invite all of your friends.

The Color of Honey

The first thing you notice about honey is its color, especially if it's been packed in a glass jar that happens to be sitting on your kitchen counter with sunlight streaming through it. The warm glow is very seductive, beckoning

you to grab a spoon and dive in. Honey can be as clear as water or as dark as molasses, though usually it's somewhere in between, a sort of amber color.

The U.S. Department of Agriculture has established standardized grades of color for honey, from the most desirable, water white, through progressively darkening grades—extra white, white, extra light amber, light amber, and finally dark amber. Although darker honeys are decidedly more flavorful than lighter ones, packers and distributors know that most consumers prefer lighter honey. I don't know why this is so. I myself have always preferred medium to darker honeys over their lighter, less flavorful cousins.

The USDA's color guide is used by beekeepers, packers, and judges to rate honey at state fairs. In international trade, the color of honey is generally specified since it's a more objective measurement than flavor.

But what causes these different shades of color? One suggestion is that a natural chemical in honey known as melanoidin reacts with pigments in tiny grains of pollen that have dissolved in the honey. Honey also darkens when it is exposed to the air, which causes oxidation. Sugars, especially fructose, react with the amino acids in honey to produce a kind of carmelization. These chemical reactions all nudge white or light amber honey into the darker grades. Honey also darkens as it ages and over time can become almost black.

The Scent of Honey

When you open a jar of fresh honey, take a moment to savor its aroma. All honeys, even the white ones, have a distinct, beguiling perfume. To fully experience the scent, warm the jar before you open it. The warmer the jar, the better, for heat releases the odorant molecules locked in the thick, viscous liquid.

The scent of honey comes from the flowers the bees have visited during their foraging. Floral scents originate in scent-releasing patches, called osmophores, on the flowers' petals. Throughout their long evolutionary history, flowering plants, needing to attract pollinators in order to survive, have become expert chemists, creating exotic perfumes that no healthy bee can resist. Thanks to these scents, the bees get to load up on nectar and pollen, the flowering plants are able to reproduce, and we can experience the exquisite aroma of honey.

The Taste of Honey

The taste of honey depends to some degree on its color. White or extra-light amber honeys have characteristically mild flavors, not as dramatic or bold as some of the darker varieties. Honey from buckwheat flowers, for example, is dark in color and has an intense flavor I happen to be fond of, though many people find it too strong. Onion honey, which bees produce from onion flowers, is also dark and flavorful but not one of my personal favorites (though it might work on an onion bagel).

If a jar is labeled "wildflower honey," it's likely to be a blend of honeys created from the nectar of many different kinds of flowers growing in one area. Or it might be a mix of honeys from many different parts of the world. Honey packers often blend cheap imported honey, usually from China or Mexico, with more expensive domestic varieties. This blending produces a standardized product, one that is consistent jar after jar, year after year, but is ultimately rather generic. Unfortunately, blending dulls the great taste you can experience with unblended honeys from a single floral source.

All honeys are not created equal. You've probably strolled past supermarket shelves or wandered among the stalls of a farmers' market and seen the honey lineup: clover, wildflower, orange blossom, mesquite, tupelo, and many more. These are the true, unblended, single-source honeys, called varietals (just as wines from a single type of grape are called varietals). Tupelo honey, for example, should only contain honey made from the nectar of tupelo blossoms. These varietals, or small-batch boutique honeys, generally cost more, but they are well worth the extra expense.

A few gourmet shops are now stocking their shelves with varietals from other countries—really exotic ones such as Tasmanian leatherwood, Eucalyptus, and the flavorful manuka honey from New Zealand. Look for them, ask for them, and experience the incredible difference they make.

Of course, the best way to appreciate the true aroma and flavor of honey is to eat it straight out of the comb. The experience is like nothing else in the world. But don't despair. Even if you don't have any beekeeping friends, all you really need to become a connoisseur is a jar of unblended honey, a spoon, and a warm slice of toast.

The National Honey Board recognizes sixty-four distinct types of honey sold in the United States or exported overseas. Some of these are blends, but most are varietals, derived from one kind of flowering plant or a few related species in the same genus. You could spend years traveling the country, visiting farmers' markets and beekeepers' roadside stands, sampling the sixty-four officially recognized honeys—which include avocado, Brazilian pepper, cat claw, thistle, pumpkin, and sumac honey, to name a few—as well as the scores of local honeys that the board doesn't classify. Or you could have a honey-tasting party.

You Are Cordially Invited . . .

Are you bored with the same old wine-and-cheese-tasting parties? Tired of drizzling designer olive oils on fresh bakery bread for your dinner guests? Then maybe it's time for a honey-tasting party.

Nothing could be simpler. You just need an interesting selection of honeys, a lot of wooden toothpicks that your guests can dip into the jars, lick, then throw away, and some moist towelettes to clean up those inevitably sticky fingers. Unlike a wine-tasting party, you won't have any fragile glasses to wash after your guests have gone home.

I'd suggest choosing five or ten honeys that vary greatly in color, aroma, and flavor. To bring out the full bouquet, gently heat the jars in a double boiler, or microwave them on a low setting for a few seconds. Warming the honey makes it flow freely (ever tried to drag a spoon through a cold jar of honey?) and releases all those luscious scent and flavor molecules, so you and your guests will be able to truly appreciate them.

You might consider making it a guessing game for your guests. Remove the labels from the honey jars (to keep track of which honey is which, write the name on a piece of paper and put it under the jar), then let everyone take a taste and see if they can identify the floral source.

For my own honey-tasting parties, I often use tupelo, blueberry, buckwheat, and orange blossom honeys—longtime favorites of mine that are available in most gourmet stores and which represent a variety of unusual flavors and colors. I might add the commonplace clover honey to serve as a benchmark. Sometimes I spice things up with a few exotics that my guests are not likely to have encountered before. These honeys might include heather honey from the British Isles, thyme honey from Greece, and perhaps green honeydew honey from France or the Black Forest region of Germany.

The important thing is to have fun and learn about the incredibly diverse world of honey. Be sure to think of the bees that made this bountiful feast possible and of the flowers that contributed their vital essences to create such an array of aroma and taste sensations.

Before you send out the invitations, however, it might be a good idea to become familiar with the honeys you intend to serve. What follows are descriptions of both common and exotic honeys from around the world.

Honeys from the United States

Common U.S. Honeys

Clover Honey *(Trifolium repens)*

Dutch clover is considered a weed when it colonizes a manicured front lawn, but when it's grown as fodder, it creates verdant pastures for both cattle and bees. The honey it yields is white, with a very mild, characteristic honey flavor and aroma that make it one of the best-selling varieties in the United States.

Buckwheat Honey *(Fagopyrum esculentum)*

Buckwheat has been cultivated since the Middle Ages as a honey plant, thanks to its nectar-rich blossoms. In the deserts of the American Southwest, native buckwheat carpets the land and provides rich forage for hungry bees. Buckwheat honey is usually dark and its flavor so intense that people either love it or hate it. Some actually liken its taste to that of molasses.

Goldenrod Honey *(Solidago spp.)*

Have you ever seen broad swaths of yellow painted across a brilliant autumn landscape? The flower responsible is goldenrod, which is also the source of a very popular honey. Because goldenrod is usually the last plant to flower before the frosts come, it's an invaluable asset for both bees and their keepers, since it yields the last honey of the season, the one that allows bees to survive the cold months ahead. The taste of goldenrod honey is not unlike that of a refreshing cup of herbal tea brewed from aromatic dried flowers such as chamomile.

Mesquite Honey *(Prosopis spp.)*

In the hot, dry climates of Arizona, Texas, and New Mexico, various species of mesquite grow in dense thickets called bosques. Beekeepers set up their apiaries near these bosques to ensure that their bees produce plenty of mesquite honey. One of the most popular varieties in the Southwest, it is light in color with a mild, delicate flavor that I personally find somewhat bland and nondescript.

Mint Honey *(Mentha spp.)*

The mint family includes peppermint, spearmint, pennyroyal, and curly mint. Honeybees are partial to mint plants because their small blossoms produce an especially sweet nectar. The light-colored honey has a mild flavor reminiscent of crushed spearmint or peppermint leaves and is perfect in iced tea.

Orange Blossom Honey *(Citrus spp.)*

Opening a jar of this honey immediately transports me back to my high school and early college days in Orange County, California. At that time, the early 1970s, the night air was perfumed by thousands and thousands of orange trees. Happily, the honey captures both this incomparably sweet, lingering scent as well as the flavor of the orange blossoms. Orange blossom honey is one of my top ten favorites, and when you try it, it will probably be on your top ten list as well.

Sunflower Honey *(Helianthus annuus and other spp.)*

Thousands of acres in the United States are planted with sunflowers, whose oil, when extracted, is widely used in cooking. Their immense "heads" track the sun as it crosses the sky and offer bees a bountiful feast of nectar and bright orange pollen. Sunflower honey has a floral aroma and taste similar to that of goldenrod honey and is light amber in color.

Rarer U.S. Honeys

Now that we've sampled the honeys you're most likely to encounter, it's time to visit the backwoods and rural lanes of America in search of rarer specialty honeys, varietals produced in much smaller quantities and not always readily available.

Avocado Honey *(Persea americana)*

Near my early-boyhood home in the rugged canyons and hills of San Diego County, California, are vast orchards of avocado trees. A native of Mexico and South America, this tree bears numerous small, greenish flowers from which bees harvest nectar. The velvety honey they produce is medium to dark in color with a flavor that packs a bit of a bite and has a piquant aftertaste that some people may not find to their liking. Fortunately, it tastes nothing like the oily avocado fruit, nor does it have much of a detectable odor—after all, who would want guacamole honey?

American Basswood Honey *(Tilia americana)*

Basswood, a member of the linden family, grows in the hardwood forests of the eastern United States. Its pale yellow flowers yield a light-colored, delicately scented honey known for its distinctive bite (probably due to alkaloids contained in the nectar).

Blueberry Honey *(Vaccinium spp.)*

The first time I tasted fresh blueberry honey from hives located near the wild blueberry barrens of Maine, I couldn't believe how delicious it was. The delicate but distinctive aroma and flavor of ripe blueberries was unmistakable. Usually, flowers and fruit produced by the same plant have different scents, but in this case they're one and the same, making blueberry honey a special treat.

Cranberry Honey *(Vaccinium macrocarpon)*

I once spent time recording the foraging behavior of honey bees in and around the cranberry bogs of Wisconsin. It's difficult to find cranberry honey outside of the relatively few cranberry-growing states (Massachusetts, Maine, New Jersey, Wisconsin, Washington, and Oregon). But if you can, it's well worth the effort. Its flavor is similar to that of cranberries but more subtle, having taken on the delicate aroma of the pink cranberry blossoms from which it is derived.

Pumpkin Honey *(Cucurbita pepo)*

The brilliant orange blossoms of male and female pumpkin flowers can be as big as a human hand and produce great quantities of nectar. But because there are relatively few blossoms per plant, even large, commercial pumpkin patches yield only small amounts of this dark-colored honey. It is typically harvested just once, in late summer or early fall, after the pumpkins have flowered and their fruits have just begun to appear. Its flavor has been described by at least one purveyor as tasting "squashy," reminiscent of the blossoms themselves.

Tupelo Honey *(Nyssa sylvatica, Nyssa aquatica)*

One of my all-time favorite honeys comes from the tupelo tree, which blooms in April and May. Actually, there are two kinds of tupelo tree, the black gum or sour gum tree and the water tupelo, both of which grow in wet areas of the Southeast, especially along the rivers of northern Florida. In the past, beekeepers used hollowed tupelo logs as hives for their colonies (shades of the Maya). Tupelo honey is much sought after in the southeastern United States. It is golden amber with a greenish tinge and greatly prized for its floral bouquet and the fact that it doesn't granulate.

Tupelo honey had a starring role in the movie *Ulee's Gold* as the crop harvested by Ulee, played by actor Peter Fonda.

Fortunately for northern and western devotees, tupelo honey can be purchased in specialty markets, from gourmet catalogs, or online anywhere in the country.

Exotic Honeys from Around the World

Honey from the Land Down Under

Australia, once part of the ancient supercontinent of Gondwanaland, broke off from that land mass and was isolated from the rest of the world for millions of years. As a result, it has produced an abundance of nectar-rich plants that exist no place else on earth.

Since their introduction by European settlers in 1822, honey bees have done a good job keeping Aussie beekeepers supplied with fat frames of honey, ripe for extracting and packaging. Let's try some of the many different varieties.

Jarrah Honey *(Eucalyptus marginata)*

Jarrah honey comes from the tallest tree in Australia, a prized hardwood that grows in the western forests of the continent.

Some have described this honey as having a rich, full flavor, reminiscent of coffee.

Patterson's Curse Honey *(Echium plantagineum)*

Patterson's Curse, also known paradoxically as Salvation Jane, is a low-growing plant whose sky-blue flowers carpet the countryside in summer and provide a wonderful crop of delicious honey. While beekeepers love the plant, Australian ranchers hate it since it can debilitate and even kill livestock that graze on its foliage. Happily, unlike the plant's leaves, the light amber, delicately flavored honey has no toxic properties. It is slow to crystallize, which gives it a long shelf life.

Rough Bark Apple Honey *(Angophora bakeri)*

This honey is not from apple trees at all but from a tall species of Eucalyptus, which flowers in December and January. Dark red in color, it has a strong, oaky flavor that leaves a slightly piquant but interesting aftertaste.

Stringybark and Messmate Honeys *(Eucalyptus laevopinea and other spp.)*

These honeys come from a group of related Eucalyptus trees native to New South Wales. They are very dark and full-bodied, with an aroma that has been compared to that of tanned leather. Fortunately, none of these honeys smells or tastes like Eucalyptus leaves, the principal ingredient of Vicks VapoRub.

Tasmanian Leatherwood Honey *(Eucryphia lucida)*

The island of Tasmania, off the southeast coast of Australia, got its first boatload of honey bees in 1831. Today, approximately two-thirds of its honey production comes from a native plant known as leatherwood. The rare leatherwood honey commands a high price and is much sought after by connoisseurs and gourmets. It is golden yellow in color, with an unusual, highly floral flavor, a piquant aroma, and a complex, lingering aftertaste. I always keep at least one jar of it in the honey cupboard of my kitchen. If you've never tasted leatherwood honey, give it a try. You won't be sorry.

Honey from the Land of the Kiwi

Like Australia, the north and south islands of New Zealand were long isolated from the rest of the world. As a result, many of their nectar-producing plants are unique to the region.

Kamahi Honey *(Weinmannia racemosa)*

The kamahi is a spreading tree with dark green, leathery leaves. A native of the temperate rainforest, it produces masses of creamy flowers that in turn yield a rich, amber-colored honey. Its full-bodied floral flavor has a buttery finish that lingers deliciously in the mouth.

Manuka Honey *(Leptospermum scoparium)*

Manuka means "tea tree" in Maori, the language of the indigenous people of New Zealand. The honey ranges in color from creamy white to dark brown and is delicious despite a hint of something slightly bitter and herbaceous. At $9,000 a ton, manuka honey may well be the world's most expensive.

The leaves of the manuka tree, which have strong antibacterial properties, have been used for centuries by the Maori as an effective treatment for wounds. Today, manuka honeys are being marketed in tubes and precoated sterile bandages as wound dressings. The types of manuka honey vary in their germ-fighting potency, but all have higher than normal activity against disease-causing microbes.

Pohutukawa Honey *(Metrosideros excelsa)*

The pohutukawa tree with its brilliant red blossoms is also known as the New Zealand Christmas tree, since it flowers during the holiday season. Its honey is my personal favorite of all the New Zealand varieties, thick and creamy white in color with a wonderful flavor that hints of butterscotch. The nectar of the pohutukawa blossoms was prescribed by Maori priests as a cure for sore throats.

Honey with a European Accent

Lavender Honey *(Lavandula angustifolia)*

Lavender honey is world renowned for its delicate floral bouquet and enchanting aroma, reminiscent of the scent of lavender in a sachet or potpourri mix. Experiencing this fine, imported honey, often used in pastries and desserts, reminds many travelers of holidays spent driving or hiking through the lavender fields of Provence, where most of the world's limited supply comes from. (Lavender is widely cultivated in Provence to supply its large perfume industry.)

Thyme Honey *(Thymus vulgaris)*

The modern Greeks are the world's largest producers of thyme honey, followed by Spain and France. While any honey can be flavored with a sprig of thyme, the real thing is made by honey bees from the nectar of thyme flowers and has an incomparably delicate flavor and aroma.

The ancient Greeks also had a taste for thyme honey. Solon, the renowned Athenian legislator, complained that nearby Mt. Hymettus was overcrowded with beekeepers and their hives and passed a law requiring that apiaries on the mountain be at least three hundred feet apart.

Heather Honey *(Calluna vulgaris)*

A low-growing plant of the heaths, mountains, and moors, heather is a familiar sight in the northern and western regions of the British Isles. From its dense clusters of violet and purple blossoms, honey bees gather nectar and ripen it into one of the world's most fragrant and enticing honeys. It varies in color from deep amber to reddish brown and can have a gel-like consistency with an interesting bittersweet aftertaste.

The Buzz at the Paris Opera

As I've often said, beekeepers are an eccentric lot—none more so than a certain French beekeeper whose activities were reported in the *New York Times* on June 26, 2003.

Jean Paucton, sixty-nine, is an urban apiculturist who keeps five weathered, wooden hives on the roof of the Palais Garnier, the venerable opera house that stands as a landmark in the center of Paris. When he visits his charges, he dons his beekeeper's hood and a pair of heavy canvas gauntlets and climbs a narrow iron ladder to a parapet only two feet wide. On one side of the parapet, the roof falls away to the teeming streets of the city below. On the other side, a skylight slopes upward, its cracked panes evidence of visitors' struggles to keep their balance while fending off Paucton's angry workers.

Monsieur Paucton, a graphic artist who spent his career as a prop man for the opera, studied beekeeping at the Jardin du Luxembourg, where a school has been teaching Parisians about hive management for 150 years. Eighteen years ago, he ordered his first hive, which was delivered to him sealed and full of bees while he was still at work. He had intended to take it to his country place, but somehow the bees never left the opera. (A precedent had already been set by a colleague who raised trout in the opera's huge underground cistern.)

Today, his five hives, overlooking the elegant avenues of Paris, house about seventy-five thousand bees, which produce a thousand pounds of honey a year. Monsieur Paucton jars and labels the honey at home, then sells it at the opera gift shop and at Fauchon, the world-famous gourmet specialty store.

According to Monsieur Paucton, his honey has a particularly intense floral flavor, thanks to the high concentration of flowering trees and shrubs that adorn the City of Light.

Bell Heather Honey *(Erica cinerea)*

Nectar collected by bees from bell heather yields a full-flavored honey, often the surprising color of port wine. Hailing from the Scottish Highlands, bell heather is one of the emblems or plant badges used by Scottish clans.

Chestnut Honey *(Castanea sativa)*

The European chestnut is especially common in northern Italy. Its honey is dark brown, similar to buckwheat honey in the United States. Like buckwheat honey, it has a taste some people love and others find quite unappealing. The flavor has been described as astringent, penetrating, and tannic, like that imparted to wine by the oak barrels in which it is aged.

Strawberry Tree Honey *(Arbutus unedo)*

In Sardinia, strawberry tree honey is known as *miele amaro di corbezzolo,* the honey beloved of honey bees. Reputed to have been a particular favorite of the Marquis de Sade, it has an almost intolerably bitter taste. (No wonder the man who gave sadism its name loved it so much.) The bright red fruit of this tree looks like strawberries from a distance. The Latin species name, *unedo,* means "I'll only eat one." Don't bother to look for this unusual honey on your local supermarket shelf—it won't be there.

Rapeseed Honey *(Brassica napus)*

Gone are the days when wildflowers blanketed hillsides and valleys across Europe. Today, much of the northern European countryside is a one-dimensional sea of yellow rapeseed, from which canola oil is made. Rapeseed honey is harvested by beekeepers in vast amounts and is often mixed with other honeys to soften its harsh, mustard leaf flavor—not one of my favorites.

But It's Green, So How Can It Be Honey?

There are a few strange honeys you may come across, called honeydew honeys, that owe their existence to nonflowering plants. These honeys begin life as a sweet, sticky residue called honeydew, excreted by aphids who feed on pine needles. The honeydew falls to the ground, where it is found by bees who transform it into a pine-green, highly flavorful honey. In France, this is called *miel de puce,* or flea honey, and it is considered a great delicacy. In the Black Forest region of Germany, honeydew from another species of pine is gathered by bees and made into a honey known as forest or fir honey.

Another type of honeydew honey is produced in New Zealand by insects living in the crevices of tree bark. These insects secrete vast quantities of processed tree sap, which the ever vigilant honey bees collect and take home to the honey factory. According to New Zealanders, honeydew honey is ideal for marinades and barbecue sauces.

In general, honeydew honeys have a very high mineral content and a pungent flavor some have compared to cough medicine.

Manna from Heaven

Many of us know that when the Israelites wandered the desert for forty years, they were sustained by manna from heaven. But no one really knows what manna actually was. One theory is that manna came from the honeydew produced by either aphids or scale insects living in the Holy Land. In Hebrew and Arabic, the word for honeydew is *man,* while *man-es-simma,* or manna, is "the honeydew that falls from the sky," which does give weight to the theory.

Chapter 9

How Sweet It Is: Cooking with Honey Throughout the Ages

The only reason for being a bee that I know of is making honey...and the only reason for making honey is so I can eat it.

—A. A. Milne, *Winnie the Pooh*

Honey can be eaten dripping and gooey straight from the comb, drizzled on a slice of hot toast, or used as an ingredient in countless dishes, from simple, down-home cooking to the sumptuous fare served at the world's finest restaurants. This chapter takes the lid off the culinary uses of honey, from ancient Anatolia and Mughal India to medieval Nuremberg, Elizabethan England, and on to modern America, with its passion for sticky barbeque sauces and sticky buns. I'll examine the role of honey in savory meat and vegetable entrees as well as in baked goods and desserts. Finally I'll provide tempting, sometimes exotic recipes from around the world, all of which use honey to make a delicious, memorable difference.

Infant Botulism and Honey

I would be remiss if I failed to warn readers about the risk of giving honey to infants, either right out of the jar or added to food and beverages. The fact is, babies under twelve months of age should never be given honey. This is because, in rare cases, honey has been found to contain spores of *Clostridium botulinum,* the bacterium that causes botulism. Infants are susceptible to this form of poisoning, though it is not a problem after their first birthday. The number of infant botulism cases reported in the United States in a typical year is about one hundred, an extremely low incidence. Some doctors believe that honey fed to babies may be related to cases of sudden infant death syndrome (crib death).

Of course, honey is completely safe for older children and adults, a healthy food to be enjoyed frequently in all its delicious varieties.

The Golden Age of Honey

Metallic gold, it turns out, wasn't the only gold that caught the fancy of Midas, the legendary king with the golden touch, who ruled in western Turkey twenty-three hundred years ago. King Midas also had a taste for golden honey. When archaeologists analyzed food residues found in his tomb, they were able to re-create the menu of his burial banquet. Mourners assuaged their grief with appetizers of goat cheese, julienned cucumbers, olive paste, and dried figs. The main course was a stew of spicy lamb and lentils, followed by a caramelized honey and fennel tart. The beverages of choice were mead (honey wine), beer, and grape wine spiced with saffron. (A traveler in modern Turkey would find many of these dishes readily available at local restaurants.)

While the chefs of old King Midas neglected to bequeath us any recipes containing honey, their Roman counterparts were more thoughtful. The recipes below are from an ancient Roman cookbook, *De Re Coquinaria,* compiled about 44 BC by Marcus Gavius Apicius, who served as culinary expert under emperors Tiberius and Augustus. Translated by Robert Maier in 1991, these recipes provide a good example of what Roman patricians ate while reclining on luxurious couches in their frescoed dining rooms, listening to the sweet strains of a lyre or perhaps poetry read aloud by an educated Greek slave.

Roman cooks used honey not only as an ingredient in many desserts and sweet-and-sour entrees but also as a preservative for meat and fruit. Favorite dishes included cheese sweetened with honey, honey omelettes, curds with honey, mushrooms sautéed in honey, and chilled white wine with honey added. The demand for honey was so great that most large Roman farms employed a full-time beekeeper, called an *apiarus,* to tend the hives.

The recipes given here are for dishes most upper-class Romans would have been familiar with.

Patina de Piris

Pear Soufflé

(Apic. 4, 2, 35)

1 kilogram (2.2 pounds) pears, peeled and cored
6 eggs, separated into whites and yolks
4 tablespoons honey
100 milliliters (3.38 fluid ounces) passum (a very sweet Roman wine sauce, something you're not likely to find on your grocer's shelf)
Olive oil
¼ teaspoon salt
½ teaspoon ground cumin
Ground pepper to taste

Preheat oven to 350°F. Cook the pears until soft, then mash and mix with lightly beaten egg yolks and the pepper, cumin, honey, passum, salt, and oil. Beat the egg whites until they come to soft peaks, then fold into pear mixture, put in a casserole, and cook approximately 30 minutes in the oven. Serve with a bit of pepper sprinkled on the top.

Dulcia Domestica

Homemade Dessert

(Apic. 7, 13, 1)

200 grams (7 ounces) fresh or dried dates
50 grams (1.7 ounces) coarsely ground pine nuts
Salt
Honey or red wine with honey

After removing the stones, stuff the dates with the ground pine nuts. Sprinkle with salt and stew in the honey or honey-sweetened red wine over low heat until the outer skins of the dates start to come off, approximately 5–10 minutes.

Sarda Ita Fit

Cooked Tuna

(Apic. 9, 10, 2)

- *500 grams (1.1 pounds) tuna fillet*
- *½ teaspoon ground pepper*
- *½ teaspoon thyme*
- *½ teaspoon oregano*
- *½ teaspoon rue*
- *150 grams (5.3 ounces) chopped dates*
- *1 tablespoon honey*
- *50 milliliters (1.7 fluid ounces) white wine*
- *2 tablespoons wine vinegar*
- *2–3 tablespoons green olive oil*
- *Garnish: 4 hard-boiled eggs, quartered*

Sauté the tuna until the flesh begins to flake and it is cooked all the way through, then mash it together with the other ingredients. Garnish with the egg quarters and serve.

A Dish Fit for a King

(Not to Mention a Queen and Her Favorite Poet)

Let's imagine that we've been invited to a sixteenth-century banquet at Hampton Court, a palace on the outskirts of London. Queen Elizabeth I is entertaining an up-and-coming young poet and playwright named William Shakespeare, whose honeyed words have captured her imagination.

The banquet is anything but a quiet literary affair, however. The palace, a Renaissance masterpiece, has been home to a number of turbulent Tudor monarchs, including Elizabeth's parents, the unpredictable

Henry VIII and his ill-fated consort Anne Boleyn. Most of the highborn guests have arrived with their retainers, who lurk vigilantly on the sidelines to make sure their masters and mistresses have all they need. Some members of the highest nobility, entangled as they are in dangerous political intrigue (for these are unsettled times), have brought along their personal tasters, to test the food for deadly ingredients that may have been secretly added to the recipes. Dogs are everywhere underfoot, gnawing the bones that, according to court etiquette, should be thrown on the floor rather than replaced on the serving tray from which they had been originally taken. Rising above the currents of boisterous conversation, the lilting strains of Renaissance music compete with clowns and acrobats leaping and tumbling through the air.

Now let's sample the rare and tempting dishes with which the long tables in the vast banquet hall are laden. This midsummer's night feast, fit for a powerful queen and a peerless playwright, might include succulent honey-glazed roasts turned slowly on a spit over a roaring fire, tiny game birds and acorn squash, lamb stew with prunes, cloves, mace, and saffron, dressed swan, pork with raisin and rosemary stuffing, salmon and figs baked in a pie, lavender biscuits, gooseberry tarts, gingerbread spiced with honey, and pears in a rich honey syrup, all washed down with large amounts of mead. The menu might also include exotic foods recently arrived in the Old World from the New World, such as tomatoes, maize, pineapple, chocolate, peanuts, hot peppers, and, just off the boat, turkey. Unfortunately for the hero of our story—and fortunately for Renaissance dentists in need of business—sugar from the West Indies is starting to make rapid inroads, replacing our beloved honey in many recipes and causing widespread tooth decay among the aristocracy, the only members of society who can afford it. Sugar, in fact, is the culprit that has blackened the teeth of Will's aging Virgin Queen. It is much more damaging to teeth than honey because it provides food for decay-causing microbes, while honey, with its antibiotic properties—see Chapter 11—kills many of those microbes.

The spices and honey used in many of the dishes not only add subtle complexities of flavor and aroma, but also mask the stale, slightly rancid taste of some of the food. The fact is, fresh ingredients are not readily available in Tudor England, and the refrigerated cases of modern supermarkets will not appear for hundreds of years.

As the evening wears on, the battalion of exhausted servers clears away the empty dishes and platters while the lute players retune their instruments and the minstrels prepare to sing. Later, the Virgin Queen, resplendent in her flaming red wig and stiffly brocaded gown, smiles with pleasure as young Will recites a few lines from his latest work in progress. No goblet is left empty, for the mead flows copiously, as it has for centuries, loosening tongues, inflaming desires, and inspiring whispered plans for midnight trysts.

If this description of Elizabeth's spread for Will has whetted your appetite for Tudor fare, give the following recipes a try in your own kitchen at home.

Gyngerbrede

Gingerbread was a popular staple throughout medieval and Renaissance Europe. The recipe below is not significantly different from those found in fifteenth- and sixteenth-century manuscripts and would have been familiar to the busy chefs at Hampton Court. Gingerbread was traditionally boiled rather than baked and was usually stamped with decorative designs. You may wish to express your own creativity with a cookie or butter press while your loaf is still warm and malleable.

*

Serves 8

1 cup honey
1 teaspoon powdered ginger
1/8 teaspoon ground cloves
1/8 teaspoon ground cinnamon
1 tablespoon anise (fennel) seeds
1 3/4 cup dry bread crumbs

Heat the honey in the top of a double boiler. Add all the spices except the anise seeds and stir to blend. Now add the bread crumbs and mix thoroughly. Cover and cook over medium heat for 15 minutes. The mixture should be thick and moist. Place the gingerbread on a large sheet of waxed paper and mold the dough into small rectangular shapes. Sprinkle the anise seeds on top and press them gently into the dough with the side of a knife. Allow to cool, then cover and refrigerate for 2 hours. Serve the gingerbread at room temperature in thin slices.

Or try this gingerbread recipe instead:

4 cups honey
1 pound unseasoned bread crumbs
1 tablespoon each ground ginger and cinnamon
1 teaspoon ground white pepper
Pinch of saffron
Whole cloves

Warm the honey and skim off any scum. Keeping the pan on very low heat, stir in the bread crumbs and spices except for the cloves. When it is a thick, well-blended mass (add more bread crumbs if necessary), remove it from the heat and allow it to cool slightly, then lay it out on a flat surface and press firmly into a square or rectangle, about ¾ inch thick. Let the gingerbread cool, then cut it into small squares to serve. Garnish each square with a whole clove.

Optional: Add a few drops of red food coloring to the mixture if, like many Elizabethans, "thou wolt have it Red."

If, after our Hampton Court banquet, we are invited to dine with the Ming emperor in China's Forbidden City, we might feast on grilled snake, bear claws, crispy king prawns served with honey-glazed walnuts, honey-glazed chicken wings (long before Buffalo, New York, got the idea), taro root with honey juice, steamed honey cakes, and, to aid the digestion, a swig of Canton ginger liqueur, rounded with honey.

Not to be outdone by his Chinese rival, Emperor Akbar of India now invites us to luncheon at the magnificent Red Fort in Delhi. Long before the Mughal conquest of India, ancient Sanskrit myths depicted the world as made up of seven concentric rings of land, separated by oceans of salt, jaggery (unrefined brown sugar), wine, ghee, milk, curds, and fresh water. All of these are still key ingredients in Indian cuisine, with the addition of honey, an important item in many Persian recipes brought to the subcontinent by the invading Mughals.

As our meal with the gracious Akbar begins, we find ourselves reclining on sumptuous pillows in a white marble pavilion inlaid with exquisite floral designs. Turbaned servants arrive bearing brass trays laden with exotic-looking dishes. After an appetizer of shrimp marinated in honey, vinegar, and spices, we turn our attention to fish baked with dill, fennel, and honey; fried dumplings dipped in honey sauce; and, for dessert, pastries stuffed with almond paste and honey and scented with rosewater, orange flower water, and saffron.

As we continue our trip through culinary history, we might have occasion to sample a breakfast of yak yogurt with honey in Tibet, grapes dipped in honey in Armenia, Persian pastries made with rosewater and honey, and snow collected in the mountains of western Iran and flavored with honey and fruit juices, the refreshing forerunner of sorbet.

Next on our gastronomic itinerary is Turkey. Considered one of the three great cuisines of the world, along with French and Chinese, Turkish cooking is known for the uniqueness of its flavors and the universality of their appeal. Turkish cuisine has influenced cooking throughout Europe, the Middle East, and Africa. It originated in Central Asia, home of the first Turkish invaders of Anatolia, and evolved over the centuries as it came into contact with cuisines of the Mediterranean cultures that the Turks conquered.

In central Turkey, the ancient city of Konya made important contributions to the Turkish diet. During the twelfth century, Konya, capital of the Seljuk Empire (the first Turkish state in Anatolia), was a renowned cultural center that attracted scholars, mystics, and poets from throughout the world. It also attracted imaginative cooks who created many of the dishes for which Konya has been famous ever since. This classic cuisine includes *böreks* (meat and vegetable dishes), *tandir kebabs* (a *tandir* is a clay oven buried in the earth), and *halva,* a sweet prepared with sesame oil, cereals, and honey.

When the Seljuk rulers were overthrown by the Ottomans in the late thirteenth century, the culinary arts in Turkey reached new heights. A visit to Topkapi Palace in Istanbul underscores the importance of fine dining to the Ottoman sultans. The huge palace kitchens were housed in several buildings under ten large domes. By the seventeenth century, some thirteen hundred workers were needed to staff the royal kitchens. Hundreds of cooks produced soups, pilafs, kebabs, fish, breads, pastries, candy, and *halva* to feed as many as ten thousand people a day.

The Arab historian Ibn Khaldun wrote, "The religion of the King, in time, becomes that of the people." The same clearly holds true for food. During the six-hundred-year reign of the Ottomans, the legacy of the imperial kitchens spread throughout their empire, influencing and refining countless regional dishes.

But enough about the sultans. Let's get a taste of something very sweet and very Turkish.

Lokmas

Honeyed Doughnuts

These light, golden fritters are drenched in honey syrup as soon as they emerge from a cauldron of hot oil and are served immediately, dusted with cinnamon. They are a favorite during religious festivals when they are offered to visitors on large brass trays.

Syrup

1 cup sugar
½ cup fresh, aromatic honey
½ cup water
1 tablespoon fresh lemon juice

Combine the sugar, honey, water, and lemon juice in a small saucepan and cook, stirring constantly, until the sugar dissolves. Turn the heat up to high and cook, uncovered and undisturbed, until the syrup reaches a temperature of 220°F on a candy thermometer, or until it thickens sufficiently to lightly coat a spoon. Pour the syrup into a bowl or heatproof pitcher and set aside to cool.

Fritters

250 grams (8 ounces) all-purpose flour
1 teaspoon salt
6 grams (0.2 ounces) dried yeast or 15 grams (0.5 ounces) fresh yeast
270 milliliters (9 fluid ounces) warm water
Pinch of sugar
300 milliliters (10.1 fluid ounces) vegetable oil, or more if necessary
1 teaspoon ground cinnamon

Sift the flour and salt into a bowl and mix in the dried yeast. Add the warm water slowly while beating with an electric mixer or balloon whisk until smooth, about 2–3 minutes. Cover with a tea towel and let the mixture rest in a warm place for 1 hour, until it has doubled its size and looks frothy.

If using fresh yeast, dissolve the yeast in about 60 milliliters (2 fluid ounces) of warm water, add the sugar to activate it, and let it stand in a warm place for about 15 minutes, until it starts to froth. (If the water is too hot, it will kill the yeast.) Empty the dissolved yeast into the middle of the sifted flour, beating continuously. Add the remaining warm water slowly, while beating, until the mixture becomes smooth, soft, and elastic. Cover with a tea towel and leave in a warm place for about 2½ hours until it rises and almost doubles in size.

Allow the oil to become very hot but not smoking in a saucepan or deep fryer, and drop a teaspoon of the mixture in it. Dip the teaspoon into a cup of cold water between each addition to prevent sticking. The lokma will puff up and rise to the surface within seconds. Turn them over and as they become pale golden—it only takes a minute—lift them out with a slotted spoon and drain them on absorbent paper. You will have around 30 lokmas. Drench them in the honey syrup, sprinkle with the cinnamon, and serve immediately.

The Venerable Honey Cake

From the palaces of the Ottoman Empire, our journey now takes us to the castles and cathedrals of medieval Germany. By the time we reach Nuremberg, we may think we have had our fill of sweets, but we really must make room for *lebkuchen,* the classical honey cake taken to new heights by the bakers of this ancient, walled city.

Of course, people had been making honey cakes for centuries. The Egyptians, believing honey to be a gift of the gods, hoped to acquire its life-giving properties by devouring it baked in cakes. Honey cakes were worn into battle as talismans and were buried with the pharaohs to accompany them to the next life. The *panis mellitus,* or honey bread, of the Romans was made of sesame flour soaked in honey after it had been cooked. Sliced and fried, it became *panis nauticus,* the sailor's biscuit. The Chinese are thought to have invented *mi-king,* a simple cake made of wheat flour and honey, around AD 900. Genghis Khan's horsemen carried *mi-king* in their saddlebags for a quick energy fix when they rode out to pillage and burn the Western world. During their conquests, the Mongols passed their taste for honey cake on to the Turks and Arabs. German pilgrims to the Holy Land acquired a passion for it and copied the recipe at home, where superstitious peasants believed it offered protection against evil spirits.

It was in a German monastery in the thirteenth century that the ancient honey cake finally evolved into *lebkuchen*—a special gingerbread made with honey from the imperial forests near Nuremberg.

Lebkuchen

Bake well in advance, as this will improve its flavor.
When stored in an airtight jar and kept in a cool place,
the lebkuchen will become softer.

450 grams (16 ounces) honey
140 grams (5 ounces) unsalted butter
450 grams (16 ounces) all-purpose flour
1 tablespoon baking soda
2 teaspoons ground ginger
1 heaping teaspoon ground cinnamon
½ teaspoon ground cloves
Pinch of ground nutmeg
Pinch of ground cardamom
Grated zest of 1 lemon (unwaxed or organic)
140 grams (5 ounces) ground almonds
110 grams (4 ounces) very finely chopped lemon and orange peel

Heat the honey and butter gently until the butter has melted. Cool. Sift the flour, baking soda, ginger, cinnamon, cloves, nutmeg, and cardamom into a mixing bowl. Add the grated lemon zest, ground almonds, and chopped peel. Pour the honey mixture over and knead it into a dough. If the mixture is too runny, add some more flour. Roll out on a floured board and cut into a variety of shapes. Bake for 8–10 minutes at 350°F. Glaze or decorate with icing if desired.

Now You're Cooking!

After our long journey through the past, we're thoroughly homesick and ready for some real American cooking. The National Honey Board suggests you use honey to bring new life to many traditional favorites. Here's a small sample of tips and recipes to give you a sense of the possibilities:

- To make your barbecue sauce more interesting, add a few spoonfuls of robust buckwheat, basswood, or sage honey.

- The next time you bake a batch of muffins, add orange blossom or clover honey to the batter.

- To glaze a roast chicken, you can't go wrong with sage or avocado honey.

- To liven up vanilla ice cream, let it soften, then spoon in some lavender or mixed wildflower honey.

- Try Eucalyptus or sage honey in hot tea. On a torpid summer day, add clover, tupelo, or orange blossom honey to a glass of iced tea.

Another way to make life sweeter is to whip up some honey butter to melt on those piping hot rolls you set out on the family dinner table.

Honey Butter

1 cup (2 sticks) unsalted butter, softened

1 cup (8 ounces) honey (use your favorite—I like orange blossom)

½ teaspoon salt

Gently beat the ingredients together in a mixing bowl until smooth, then chill and enjoy.

Variations: Make herbed honey butter by mixing in 2 teaspoons fresh or dried rosemary, thyme, or lavender. For lemon honey butter, just add 2 teaspoons fresh lemon juice; I also like to add 1–2 teaspoons of grated lemon zest.

From honey butter, it's a short, easy step to . . .

Honey Mustard

¼ cup (2 ounces) mild honey, such as clover

¼ cup Dijon mustard

½ cup mayonnaise

Place the honey and mustard in a mixing bowl and whisk until smooth. Then whisk in the mayonnaise and chill in the refrigerator. When it's ready, you'll have a tangy dipping sauce, perfect for raw vegetables.

Variation: The honey mustard dipping sauce can be easily transformed into a zesty salad dressing with the addition of balsamic vinegar, a small amount of olive oil, black pepper, and maybe some chopped chives.

I've even been known to add a few ground chiltepins (small, fiery chilies the size of a garden pea and the wild progenitor of all peppers) to create a nice hot honey mustard to bring sandwiches to life. In fact, I have a chiltepin plant growing right outside my kitchen window and pluck its fruit often. My daughters are right—I'm a bona fide chili head from Tucson and proud of it. If you can't stand fire, stay out of my kitchen. (I have to admit, though, that chiltepins mixed with ice cream, a recipe concocted by my friend Gary Paul Nabhan, is an acquired taste. However, honey and vanilla extract added to ice cream is a wonderful treat.)

Tips for Cooking with Honey

Here are some tips to keep in mind when cooking with honey. First, honey should always be stored at room temperature, never in the refrigerator or freezer. Keep it in your cupboard or pantry. It's natural for honey to change in character after a few months on the shelf. It may become cloudy or start to granulate (tiny crystals form, turning the liquid honey into a semisolid). Honey that has clouded or granulated can easily be restored to its original golden color and texture by applying a bit of heat. Be careful, though. If heated too much or too fast, honey can scorch or burn, losing its wonderful aroma. I usually warm it in a double boiler, submerging half to three-quarters of the uncovered jar in water. Stir the honey frequently and the cloudiness and crystals will disappear.

If you are in a hurry, uncap the honey jar and place it in a microwave oven. Make sure the setting is on low or medium at most. Stop and check the honey every thirty seconds or so. Whatever you do, don't boil it. You want it just hot enough to dissolve the crystals.

Cooking with honey can be a sticky business. To avoid a mess, use a traditional wooden or plastic honey dipper. It looks like a spoon but has little flanges to prevent the honey from dripping. Better yet, store your honey in a plastic squeeze container. There's almost no dripping with this method.

Measuring honey in a glass or plastic measuring cup can present a problem. To make sure the honey pours out of the cup easily, apply vegetable oil to the inside of the cup. If your recipe calls for both oil and honey, just measure the oil first and then measure the honey in the same cup. Leaving the honey out at room temperature or gently warming it makes it easier to pour and mix with other ingredients.

Cooking with honey can really make a difference in your enjoyment of many foods. It adds to and brings out the flavor of the other ingredients it's mixed with. It keeps breads and cakes especially moist and flavorful and even adds to their normal shelf life. (That's because honey attracts and absorbs moisture.) Honey can also add a rich, golden color to the crusts of pies and tarts.

Sugar and honey aren't created equal. Honey is 80 percent sugar and 20 percent water, while cane and beet sugars are 100 percent sugar. When substituting honey for table sugar in a recipe, you will need to make a few small but important adjustments. As a general rule, when replacing sugar with honey, use half the amount of honey as sugar called for in the recipe. Because honey is 20 percent water, you need to reduce the amount of liquid (water, milk, fruit juice, etc.) in your recipe by 1 tablespoon for each 4 tablespoons of honey. It's also a good idea to add ½ teaspoon of baking soda to the recipe for each cup of honey used. When baking your sweet sensation, remember to reduce the oven temperature by 25°F. This will prevent overbrowning.

National Honey Board Recipes

The National Honey Board has collected scores of recipes that call for honey. I've picked a few of my tried-and-true favorites to share with you. To explore further, just log on to the Honey Board site at www.honey.com/recipes.

Teriyaki Honey Chicken

Makes 4 servings

1 medium-sized fryer chicken, cut up
½ cup honey
½ cup soy sauce
2 medium garlic cloves, crushed
¼ cup dry sherry or water
1 teaspoon grated fresh ginger or 2 teaspoons ground ginger

Place the chicken in a large glass baking dish. Combine the remaining ingredients in a small bowl and pour over the chicken, turning to coat all sides. Cover the dish with plastic wrap and marinate in the refrigerator for at least 6 hours, turning two or three times.

Remove the chicken from the marinade, reserving the liquid. Arrange on a rack over a foil-lined broiler pan. Cover with foil and bake at 350°F for 30 minutes. Uncover and brush with the marinade. Bake, uncovered, for 30–45 minutes or until done, brushing occasionally with the marinade.

Honey Garlic Pork Chops

Makes 4 servings

1 tablespoon dry sherry
2 garlic cloves, minced
¼ cup (2 ounces) honey
¼ cup lemon juice
2 tablespoons soy sauce
Four 4-ounce boneless center-cut pork chops, lean

Combine all the ingredients except the pork chops in a small bowl. Place the pork in a shallow baking dish and pour the marinade over it. Cover and refrigerate 4 hours or overnight. Remove the pork from the marinade; heat the remaining marinade in a small saucepan over medium heat until it simmers. Broil the pork 4 to 6 inches from the heat source for 12 to 15 minutes, turning once during cooking and basting frequently with the marinade. May also be grilled.

Pacific Rim Honey-Grilled Fish

Makes 4 servings

¼ cup honey
¼ onion, chopped
2 tablespoons lime juice
2 tablespoons soy sauce
2 tablespoons hoisin sauce
2 garlic cloves, minced
1 whole jalapeño pepper, seeded and minced
1 teaspoon minced fresh ginger
1 pound swordfish steak or other firm-fleshed fish

Combine all of the ingredients except the fish and mix well. Marinate the fish in the honey mixture for at least 1 hour in the refrigerator. Barbecue or broil the fish, allowing 10 minutes per inch of thickness, or until the fish flakes when tested with a fork.

Southwestern Lasagna

Makes 4 servings

1 tablespoon vegetable oil
1 medium onion, thinly sliced
1 garlic clove, finely chopped
1 tablespoon chili powder
1 tablespoon paprika
¾ cup water
1 can (6 ounces) tomato paste
¼ cup honey
¼ cup fresh lime juice
1 15-ounce can black beans, undrained
1 can (12 ounces) whole-kernel corn
6 medium corn tortillas, cut in quarters
1 package (15 ounces) part-skim ricotta cheese
½ cup (2 ounces) shredded Monterey Jack cheese

Heat the oil in a medium saucepan over medium-high heat until hot; add the onions and garlic and cook, stirring frequently, for 3 to 5 minutes, or until the onion is tender. Add the chili powder and paprika; cook, stirring, for 1 minute. Stir in the water, tomato paste, honey, and lime juice until well mixed. Then stir in the black beans and corn. Bring to a boil; reduce the heat and simmer for 5 minutes.

Spoon one-third of the sauce into a 1½-quart rectangular baking pan; arrange half of the tortilla quarters evenly over the sauce in the pan. Spread with half the ricotta cheese. Make a similar layer with another third of the sauce, the remaining tortillas, and the ricotta cheese. Spread the remaining sauce evenly over the top of the lasagna; sprinkle evenly with the shredded cheese. Bake at 350°F for 20–25 minutes, or until heated.

Butternut Squash Soup

Makes 6 servings

2 tablespoons butter or margarine
1 onion, chopped
2 garlic cloves, minced
3 carrots, diced
2 celery stalks, diced
1 potato, peeled and diced
1 medium-sized butternut squash, peeled, seeded, and diced
3 cans (14.5 ounces each) chicken broth
½ cup honey
Salt and pepper to taste

In a large pot, melt the butter over medium heat. Stir in the onion and garlic. Cook, stirring, until lightly browned, about 5 minutes. Stir in the carrots and celery. Cook, stirring, until tender, about 5 minutes. Stir in the potato, squash, chicken broth, and honey. Bring the mixture to a boil; reduce the heat and simmer 30–45 minutes, or until the vegetables are tender. Remove from the heat and cool slightly. Working in small batches, transfer the mixture to a blender or food processor; process until smooth. Return the pureed soup to the pot and season to taste with salt and pepper. Heat and serve.

And for Dessert, You May Want to Try . . .

This recipe is also courtesy of the National Honey Board.

Honey Lemon Squares

Makes 12 servings

½ cup butter or margarine, softened
¼ cup confectioners' sugar
1 cup plus 1 tablespoon flour, divided
¾ cup honey
½ cup lemon zest
3 eggs
1 teaspoon grated lemon peel
½ teaspoon baking powder

In a medium bowl, cream the butter and sugar until light and fluffy. Add 1 cup flour and mix until combined. Press the mixture evenly into the bottom of a 9-inch-square pan. Bake at 350°F for 20 minutes, or until lightly browned. Meanwhile, in a medium bowl, whisk together the remaining ingredients until thoroughly blended. Pour over the baked crust and bake 20 minutes more, until the filling has set. Cool in the pan and cut into squares to serve.

Now, from the Kitchen of Stephen Buchmann

I've always loved cooking, whether for myself, my daughters, or friends, neighbors, and out-of-town colleagues. To me, paradise is barbecuing outside as the sun sets behind the jagged peaks of the Tucson mountains.

Some call it work—the shopping, washing, cutting, and preparing, not to mention the piles of dirty dishes to clean up—but for me, cooking is wonderfully therapeutic and stimulating. Hardly anything is more satisfying than combining ingredients in novel ways to create tasty, healthful meals, especially when one of those ingredients is honey.

Whether used in a sweet dish or a savory one, honey adds a new dimension to even the most ordinary foods, reminiscent of bees, flowers, and blissfully warm, sunny days. To liven up my inventions, I use desert honeys harvested from my own hives as well as more exotic ones from distant lands. Here are a couple of my tried-and-true favorites.

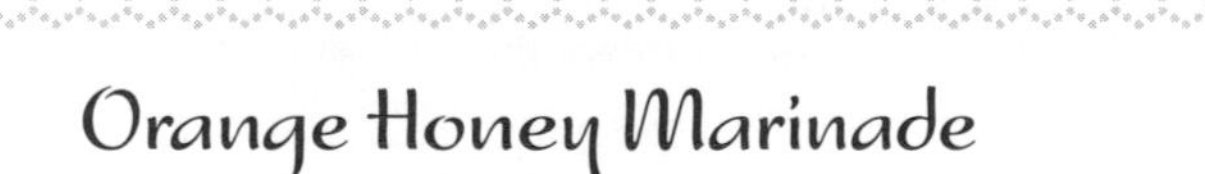

Orange Honey Marinade

This one couldn't be easier.

1 cup (8 ounces) orange blossom honey
2 tablespoons freshly grated zest from 1 orange
1 tablespoon grated fresh ginger

Blend the ingredients in a mixing bowl, add your choice of meat (such as Cornish game hen, chicken, or turkey), and marinate for 1 hour or overnight in the refrigerator, letting the flavors blend. Use the remaining sauce to baste the meat while cooking on the barbecue or roasting in the oven.

Buzzman's Zesty Honey Chicken Tenders

Makes 8 servings

½ to 1 cup clover or wildflower honey
¼ to ½ cup water
2 tablespoons freshly squeezed lime juice
1 tablespoon grated fresh ginger
Freshly grated zest from ½ orange
¼ teaspoon powdered cumin
1 tablespoon dried Mexican (not European) oregano, crushed, then powdered using your fingers or a mortar and pestle
2–4 medium garlic cloves, minced or pressed
Salt and pepper to taste
½ diced fresh jalapeño pepper or 2 dried chiltepins, crushed (optional)
2 pounds chicken breast, skinned, boned, and cut into narrow strips about ½ to 1 inch wide

Mix all of the ingredients except the chicken strips until smooth. Pour into a large zip-top plastic bag. Add the chicken strips to the bag and seal. Shake thoroughly and chill in the refrigerator for at least 1 hour prior to cooking; overnight is even better. Remove the marinated chicken strips and drain, reserving the marinade. Cook the chicken on the barbecue or over medium heat on the stove until done, adding the reserved marinade to the chicken as required to keep it moist. Serve over rice or with your favorite vegetable.

Our gastronomic tour has taken us from the kitchens of antiquity to the banquet halls of Hampton Court palace and the Forbidden City of Imperial China, right to my own backyard barbecue in Tucson, Arizona. Along the way, we've discovered that honey works its magic in baked goods, desserts, savory meat and fish dishes, and countless sauces and marinades.

Cooking with honey reinforces that ancient bond between bees (and the golden treasure they produce) and us. So try some of the recipes I've given here and enjoy a feast of thanks with family and friends, homage to the hardworking honey bee.

Chapter 10

Mead: The Honey that Goes to Your Head

Who am I?

I am valuable to men, found far and wide,
brought from groves and mountain slopes,
from hills and dales. By day wings
carried me aloft, conveyed me skillfully
under a roof's cover. Afterwards men
bathed me in a vat. Now I am a binder
and a flogger, quickly throw a young
man to the ground and sometimes an old peasant.
He who grapples with me and contends
against my strength soon finds
that he must seek out the earth with his back
unless he first gives up his folly.
Robbed of strength yet strong in speech,
bereft of might, he has no control of his mind,
of his feet or hands. Find out what I'm called,
who thus binds foolish young men on the earth
after the fight by the light of day.

Riddle #25, Exeter Book,
England, circa AD *950*

(The answer, if you haven't guessed, is mead.)

Water sweetened with honey and allowed to ferment was the *melikraton* of the Greeks, the *aqua mulsa* of the Romans, the *meda* of medieval Prussia, the *aguamiel*

of Spain, and the *tschemiga* that warmed the Russian soul long before anyone thought of vodka. We call it mead, the most popular fermented drink of the ancients, and its history is a long and colorful one.

There are many who insist that mead, not beer from hops, is the world's oldest alcoholic beverage. Mead's claim to primacy seems credible, since cultivated grains, such as the wheat and barley from which beer is brewed, first appeared ten thousand years ago in the Fertile Crescent, by which time honey hunting was already a well-established practice. Perhaps the first mead was produced when prehistoric honey hunters left a gourd of honey uncovered during a rainstorm and came back to find a potent surprise awaiting them. Wild yeasts are everywhere, and once honey is diluted with water, they cause the mixture to quickly ferment into mead.

Like honey, mead played an important role not only in the daily life of many ancient peoples but also in their mythologies and religious ceremonies.

The Love of Mead

Throughout history, mead was consumed in prodigious quantities by gods, saints, kings, queens, warriors, and commoners.

India

In the *Rig Veda,* a collection of four-thousand-year-old Sanskrit hymns that formed part of the religious canon of the Indo-Aryan peoples, two of the most important gods, Krishna and Indra, were referred to as *madhava,* "honey-born," and were symbolized by the bee. With honey such a key ingredient in their divine mythologies, it's not surprising that the Aryans believed there was a natural source of mead in heaven: *In wide-striding Vishnu's highest footsteps there is a spring of mead.* To find instant happiness, the Aryan gods had only to visit the spring and drink their fill.

Scandinavia

In Norse mythology, mead helped the gods unwind after long,

arduous days spent harassing their human underlings. Many Norse legends report tales of enterprising gods and goddesses using mead to enhance their powers as well as to relax. Like literary masters throughout history, Odin, the king of the Viking gods, found that alcohol (in the form of mead) heightened his gifts of poetry and composition. Bragi, son of Odin and heir to his literary mantle, was the god of eloquence and the patron of poets. Ancient runes carved on his tongue inspired humans to write verse so that they too might taste the "mead of poetry."

The Valkyries (whom Wagner was to make famous throughout the operatic world centuries later) were handmaidens of Odin whose task was to choose the heroes destined to die in battle. After the bloody carnage, the maidens conducted the fallen warriors to Valhalla, the Hall of the Slain Warriors. During feasts prepared for the heroes as compensation for their ultimate sacrifice, the Valkyries served them mead, which flowed copiously and conveniently from the teats of Odin's she-goat, Heidrun.

The Near East

Evidence indicates that one of the earliest royal devotees of mead was none other than the fabled King Midas, who lived around 700 BC and ruled the land of Phrygia in what is now modern Turkey. In 1981, after more than twenty years of exploratory digging in about a hundred burial sites, an archaeological team from the University of Pennsylvania finally struck pay dirt. Buried deep in a huge mound that looms over the village of Yassihoyuk, southwest of Ankara, they discovered the tomb of the legendary ruler. As we've already seen, the tomb contained traces of royal dishes prepared with honey, but that's not all. It seems that the rapacious monarch had a lust for gold in its liquid form—mead. The tomb's contents included a number of lion- and ram-headed buckets, two large cauldrons, and a hundred bronze drinking bowls. Laboratory findings suggest that the buckets were probably used to transfer mead from the cauldrons into smaller jugs, which servers then emptied into the drinking bowls.

The Midas tomb team, led by Dr. Patrick McGovern, used high-tech infrared and mass spectroscopy to study the residue at the bottom of the bronze drinking vessels. The remains of herbs, pollen, grains, and other materials indicated that Midas's golden drink was made from yellow muscat grapes, toasted barley malt, thyme honey, and saffron—a recipe that, if followed today, would yield a fine, dry honey wine.

Greece

Like Odin and Midas, the Greek gods and goddesses were all partial to mead. In fact, Dionysus was the god of mead long before he became the celebrity spokesgod for grape wine. The love goddess, Aphrodite, was so fond of mead that amorous ladies of the Greek upper class, anxious to find the Adonis of their dreams, made offerings of honey wine in order to enlist her help.

It was said that the oracles of Delphi derived much of their inspired foresight from generous drafts of mead. In fact, mead was so important to successful prophecy that the virginal maidens were often referred to as the Melissai, from the Greek root *mel,* meaning "honey." As an interesting aside, the priestesses at Delphi as well as those at the Temple of Artemis in the Greek city of Ephesus were called bees, perhaps because of their mead-induced forecasting as well as their busy devotion to duty.

Although grape wines were readily available in ancient Greece, mead was the preferred alcoholic beverage. After a new batch was made, it was set aside to age until the time had come to celebrate the Dionysian rites, which took place twice a year. During these rites, men and women clad in animal skins climbed to the top of local mountains, where they sacrificed animals to the gods, drank vast quantities of mead, and made love while under the influence.

According to some accounts, mead almost caused the Greek philosopher Hippocleides to sabotage his marriage on his wedding night. Having overindulged in honey wine, perhaps the result of too many long-winded toasts, he shed his clothes, stood on his head in the middle of the wedding feast, and sang a highly inappropriate song to his by now thoroughly blushing bride. His father was not pleased nor were the authorities, and Hippocleides was roundly censured for his outrageous behavior.

Rome

Mead was also important to the Romans, who made regular offerings of honey wine to their revelry-loving gods and goddesses. On a more down-to-earth note, the patersfamilias of many Roman families offered mead as a sign of devotion to various household deities, especially the Penates, who presided over the all-important kitchen and food cupboards.

Columnella, the Hispano-Roman naturalist, gives a recipe for mead in *De Rustica,* his agricultural treatise written about AD 60. Pollio Romulus, a prominent patrician, wrote to his emperor, Julius Caesar, extolling the benefits of mead. He attributed the fact that he had sired many children and was still enjoying an active sex life at the age of one hundred to his lifelong habit of drinking mead, copiously and often. This is but one of many Roman accounts linking mead to a strong libido and sexual prowess lasting well into old age.

Germany

While Greeks and Romans sipped mead out of finely wrought goblets, their barbarian neighbors to the north imbibed theirs from flagons, wild cattle horns, and huge cauldrons that held enough to inebriate an entire roomful of banqueters.

A dig at the Iron Age Hochdorf tomb in what is now Germany brought to light several artifacts of interest to both mead drinkers and historians. An enormous bronze cauldron found in the tomb and manufactured in western Greece was probably used to serve mead. Modern estimates put its volume at 600 pints or 350 liters. That's a lot of mead. By the time archaeologist Jorg Biel excavated the tomb in 1978–79, the mead had solidified into a brown, cakelike deposit in the bottom of the cauldron, but pollen analysis confirmed that the substance had contained honey. The powerful chieftain buried in the tomb was laid out on a bronze couch. He is estimated to have died in his mid-forties, an advanced age for the period—perhaps due to his taste for mead. The personal adornments buried with him included a birch-bark hat, a gold necklace, and a gold bracelet. His belt, his dagger, and even his shoes were covered in gold foil. Along the south wall of the tomb hung eight magnificent drinking horns, all highly decorated and rimmed with gold. Above the chieftain's head was another drinking horn, capable of holding ten pints of mead. (Drinking horns, by the way, were used in some parts of Europe until well into the nineteenth century.)

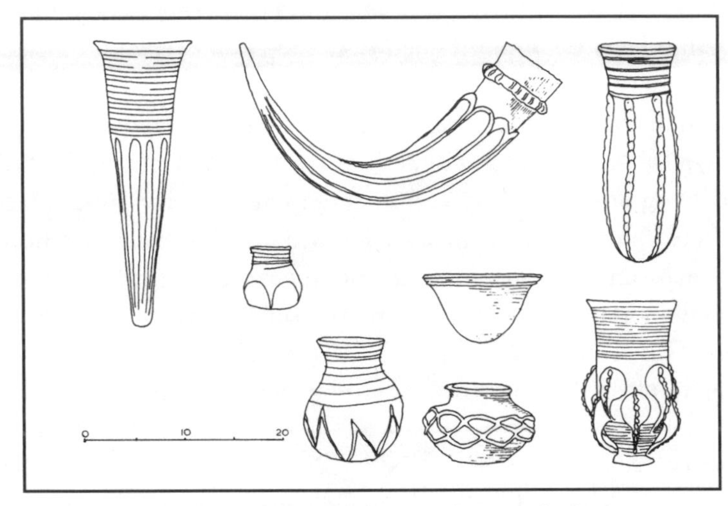

Medieval glass drinking horns and various beakers, used for consuming mead in bygone days.

Far bigger than the drinking horns and cauldrons found in the Hochdorf tomb was the krater or jar discovered in the tomb of a Burgundian princess, who died around 450 BC. Used to mix and serve enormous quantities of mead, this monumental krater, cast in bronze with gorgon-head handles and an exquisite frieze, had been imported from Greece and is one of the most spectacular pieces of classical Greek metalwork to survive anywhere.

The Celtic Lands

Drinking mead wasn't just about debauchery—and wasn't just for Norse gods, gold-obsessed kings, and Burgundian princesses. The Celtic tribes, who had spread throughout Europe before the rise of the Roman Empire and included the Irish, the Welsh, and the ancient Gauls, believed that their beloved honey wine conferred courage before battle and strength during it. After the battle was won, mead was the beverage of choice for the obligatory thanksgiving feasting. And when the party was over, mead still had a role to play, for it was believed to be a powerful aphrodisiac.

Inebriation among the pre-Christian Celts was a communal experience which closed deals and sealed alliances. It played an important part in the festival of Samhain, the Celtic new year, which began on November first. The rite of intoxication, like honey itself, was linked to fertility, bountiful harvests, and success.

Occasionally, mead got involved in politics. When the early Irish, part of the Celtic family, dispatched their disgraced kings to the next world, they did so by drowning them in a vat of mead and setting their palaces ablaze. Presumably death by mead was not a bad way to go. Mead was as important to the pagan Irish as whiskey and Guinness are to their descendants.

Drunkenness was not condemned in the ancient world, for it made men feel like gods. After all, mead was the drink of the immortals. None of the Celtic deities denied themselves a healthy swig from time to time. And mead seems to have remained in good repute even in Christian times. In post–St. Patrick Ireland, the good St. Finian lived on bread and water all week long but feasted on salmon chased down with a bottle of mead on Sundays. Another Irish saint, Brigit, was the patroness of amateur mead makers. When the king of Leinster dropped by for cocktails, she found she had run out of mead, a calamity for anyone. Brigit saved the day by using her powers to turn a large vat of water into the incomparable honey wine.

England

By the time we reach seventh-century England, every Anglo-Saxon castle boasted a mead hall, where mead was served to noble guests with great ceremony. Often the lady of the castle herself, adorned with rings and flashing gems, would make the rounds of the hall, offering the mead cup to young and old alike.

About a thousand years later, mead was still being served in the castles and palaces of England. Queen Elizabeth I loved her mead, although her favorite recipe, which called for spices, syrup, grape wine, and honey, produced a sickly-sweet concoction that would not appeal to the un-Tudored palates of modern Chablis drinkers.

Another great believer in mead, Sir Kenelm Digbie, who was born the year Elizabeth died, wrote a book called *The Closet of the Eminently Learned* (and modest, I might add) *Sir Kenelm Digbie, Kt. Opened.* In the book, he stated:

> *Meathe [mead] is singularly good for consumption, stone, gravel, weak-sight, and many more things. A chief burgomeister of Antwerp used for many years to drink no other drink than this and though he were an old man, he was of extraordinary vigor, had always a great appetite, good digestion and had every year a child.*

The Muslim World

The medicinal properties of mead were also appreciated throughout the Muslim world. Although the Quran clearly forbids the consumption of wine and other alcoholic beverages, Muslim religious leaders, in the interest of promoting good health among their followers, made an exception in the case of mead. Mead is mentioned in the ninth-century *Aqrabadhin of Al-Kindi,* a highly regarded medical formulary compiled by Abu Yusuf Ya'qub ibn Ishaq al-Kindi. The author was a respected Baghdad scholar who wrote about sorcery, astrology, and cooking as well as medicine. The kind of mead al-Kindi describes in his formulary is actually what is called sack pyment—grape juice to which honey is added and allowed to ferment. The term *sack* refers to mead that contains roughly eight pounds of honey to five gallons of grape juice.

The Bambara, a Muslim ethnic group living in the West African nation of Mali, still consider mead the drink of good health, knowledge, and truth. (Like a honeycomb, they say, truth has neither a wrong side nor a right side and is the sweetest thing in the world.) Following the precedent set by religious lawmakers in al-Kindi's time, the Bambara believe the therapeutic and spiritual value of mead exempts it from the Quran's prohibition of alcoholic spirits.

The Decline of Mead

There are many factors that contributed to the decline of mead. It probably started with the gradual debasement of the art of mead making, beginning in the fourteenth century. As the science of winemaking improved, local aristocracies turned more and more to the fashionable new vintages, a sure sign of sophistication, and the demand for high-quality mead decreased. With business falling off, many skilled mead makers took up other trades, leaving the brewing of mead to untrained peasants. Before long, old recipes were lost or forgotten. Attempts to mask the inferior quality of the new mead by adding spices and herbs fooled no one with a discerning palate.

Still, despite the defection of the gentry and the decline in its overall quality, mead remained a popular and widely available drink among the lower classes and, oddly enough, at the Tudor court in England, where, as we have seen, a very sweet type of mead was the favorite of Queen Elizabeth herself. (Elizabeth's partiality may have been due to her Tudor genes, for the turbulent dynasty originated in Wales, a region which remained loyal to mead long after the rest of Britain had switched to wine.)

During the seventeenth century, mead was dealt another serious blow. Honey supplies began to shrink as forests and flower-filled meadows fell under the plow or disappeared beneath growing cities and towns. The loss of foraging opportunities for bees led to a decrease in their production of honey. This drove up the cost of honey and therefore of honey wine. At the same time, the spread of vineyards throughout continental Europe ensured the availability of ever greater quantities of grape wine at ever lower prices—a free-market condition that further eroded the popularity of mead.

The real coup de grâce came in the eighteenth century. Until then, sugar imported from the East and called Indian salt was a luxury so expensive only the aristocracy could afford it. But the colonization of the West Indies had led to the establishment of vast sugarcane plantations fueled by slave labor, which produced large supplies of sugar at prices nearly everyone could manage. Honey, increasingly hard to find and increasingly expensive, just couldn't compete, and the production of mead continued to fall. By the middle of the century, Europe's hankerings for sweet alcoholic beverages were being satisfied by sherry from Spain and port from Portugal—and the golden age of mead was over.

Making Mead Yourself

Although mead drinking and mead making have long been out of fashion, there are pockets of loyalists, mostly amateur, who continue to make a surprising variety of honey wines, from the old English *meodu* and the spiced Welsh drink called *meddeglyn* to the Arab *nabidh*, the Ethiopian *t'ej*, and the Greek *ydromeli*.

Even if you can't lay your hands on any of these, it's possible to discover the pleasures and benefits of mead making for yourself. Good ingredients and carefully watched fermentation are the keys to success. I've made and enjoyed plenty of mead, particularly during my days as a grad student. A number of books and Web sites can tell you how to do it. I recommend *Making Mead* by Bryan Acton and Peter Duncan. The Web site of the National Honey Board is also an excellent resource.

Traditional mead can be made with any kind of honey. A light honey, such as clover or tupelo honey, will produce a lightly flavored mead. Orange blossom or lavender honey results in a deeply flavorful, aromatic honey wine. Dark, full-bodied honeys produce finished meads with even stronger flavors. Just as there are both amber beers and heavy, almost black ales and stouts, mead can be brewed as light or as dark as you please. Meads made from light honey can be fermented to produce a dry, light wine similar to sauterne, with a typical alcohol concentration of 10–11 percent. They can also be refermented in the bottle to yield excellent sparkling meads that resemble champagne. Light meads can also be used in the production of distilled spirits, including sherry and brandy.

The simplest way to make mead is to follow the recipe provided by the National Honey Board. It requires about two and a half pounds of honey, a gallon of water, and a little help from the living microorganisms we call yeasts, which turn most of the sugars in honey into ethanol and carbon dioxide, thereby causing it to ferment. Many mead makers advise adding yeast nutrients (vitamins and minerals) to the must (the starting mixture of honey and water). This is a good idea, since it speeds up the fermentation process without sacrificing the quality of the mead. When spices, herbs, or fruit juices are added to the must, fermentation bubbles along so quickly that the extra yeast nutrients aren't necessary.

Once you've made your mead, enjoy it in moderation. No cauldrons or kraters, please.

Chapter 11

Good for What Ails You

If the dew is warmed by the rays of the sun, not honey but drugs are produced, heavenly gifts for the eyes, for ulcers and the internal organs.

—Pliny the Elder, *Natural History*

Actress Julie Andrews, starring as Mary Poppins in the popular movie of the same name, reminded us that "a spoonful of sugar helps the medicine go down." I would go two steps further: not only will honey make the medicine go down better than sugar, it actually *is* medicine. From our earliest history down to the present day, people have been fortifying themselves and curing diseases and infections with honey. While some of its medicinal uses have fallen out of favor and been discredited as mere fads, others have actually been shown to produce genuine therapeutic results. In fact, modern scientists are actively studying a wide range of new medicinal applications for honey.

In writing about honey as a treatment for human maladies, I am not suggesting that you move it from your kitchen pantry to your medicine cabinet. After all, my doctorate is in entomology, not medical science. But I do

find it fascinating to see how honey has been used as a medicine over time and across cultures and to realize that our age-old bond with bees doesn't just sweeten our lives but heals our bodies as well.

The Germ-Killing Properties of Honey

There is no doubt that honey is an effective antibiotic. Why? The answer is short and sweet: sugar. Honey is roughly 80 percent sugars and 20 percent water (with a few minor ingredients thrown in for good measure). The sugars are what make honey such a quick energy booster. But can they really kill most of the rogue's gallery of bacteria and fungi that infect our world? Well, yes, they can. It's all about something called osmotic pressure. Osmosis is the movement of a solvent through a semipermeable membrane into a solution of higher concentration. The high sugar content of honey, along with its natural acidity, creates a very unpleasant habitat for single-celled microbes. These troublesome little creatures come encased in a thin, semipermeable membrane through which water is able to pass. Think of a bacterium as a water balloon sitting in honey. Thanks to the osmotic pressure exerted by the highly concentrated sugars, the water molecules in the bacterium, essential to its survival, are drawn through the thin membrane into the honey. With the help of a high-powered microscope, you can actually see the bacterium shrinking before your eyes. The story doesn't have a happy ending for the dehydrated microbes, which, unable to withstand the osmotic pressure, simply shrivel up and die.

Osmotic pressure isn't the only reason honey is an effective antibiotic. It also kills bacteria because it contains hydrogen peroxide. That's right, the same hydrogen peroxide that's in the brown plastic bottle in your medicine chest. Glucose oxidase is an enzyme secreted by bees when they convert nectar into honey. In the presence of oxygen, the enzyme splits glucose molecules into water and hydrogen peroxide. Full-strength honey has very low amounts of hydrogen peroxide and not much active glucose oxidase. When it's diluted, however, a huge increase in enzyme activity occurs. This makes honey a slow-release antiseptic, one that does not damage tissue as other antiseptics sometimes can. Modern laboratories around the world have tested the killing power of honey against a number of bacterial pathogens, with deadly results to the infection-causing bugs. While pharmaceutical companies have yet to produce honeyed antibiotics, bandages impregnated with honey are already used in Europe, Australia, and New Zealand to help prevent wounds from becoming infected.

Although the ancients didn't have microscopes to observe the potent antibacterial properties of honey, through trial and error they learned that if they applied honey and grease, or honey alone, to wounds, they healed faster and didn't get infected. Amazing.

Healing with Honey Throughout the Ages

We will never know the exact sequence of events, but early in our past, we came to rely on honey not only as a food but also as a medicine with a wide range of benefits.

One prescription for the wonder drug of the ancients is found in the *Vedas*, the sacred Hindu books compiled between 1500 and 500 BC: *Let one take honey . . . to beautify his appearance, develop his brain, and strengthen his body.* We have to wait until about 2000 BC for the first written instructions for the preparation of a honeyed healing agent: *Grind to a powder river dust . . . then knead it in water and honey and let plain oil and hot cedar oil be spread over it.* Scratched in cuneiform onto the surface of a baked clay tablet, this Sumerian prescription has been interpreted by scholars as the recipe for an unguent to treat disorders of the ears and eyes. Other Sumerian writings from that period refer to honey mixed with butter or other animal fats to form a greasy paste used to heal pierced earlobes and surgical incisions. Mixed with grains or herbs, honey was also applied to the countless cuts and scrapes incurred in the course of a typical ancient workday.

The early Egyptians also appreciated honey's healing properties, raiding their cylindrical clay hives to treat a host of medical complaints. Their honey Rx's appeared not on the clay tablets used by the Sumerians but on papyrus, a durable parchment made from reeds that grew in the extensive marshes along the banks of the Nile. A large number of these papyrus records have been recovered from tombs and deciphered by modern Egyptologists. One of the most famous is the Ebers Papyrus, among the oldest medical texts to have been found in Egypt, dating from about 1550 BC. It contains recipes for medicinal preparations as well as instructions for physicians explaining how to use them. Out of 700 formulas, 147 call for honey as one of the principal healing agents. To cure baldness, physicians were advised to concoct a mixture of ground red ocher, powdered alabaster, and honey. Other papyri of the same period instruct healers to use honey to treat skin conditions as well as infections that set in after routine surgeries, including the circumcision of male infants.

The Edwin Smith Papyrus (named for the man who purchased it along with the Ebers Papyrus in 1866) details forty-eight possible uses for honey as a healer, two of which we've included here.

> Case Two: Instructions concerning a gaping wound in the head, penetrating to the bone
>
> *To examine a man having a gaping wound in the head penetrating to the bone, you should lay your hand on the wound and palpitate it. If you find the skull is uninjured, not having a perforation in it, you should bind fresh meat to the wound with two strips of linen to draw the wound together, then treat it with grease and honey every day thereafter until he recovers.*

The Egyptian physician who wrote this case knew what he was talking about. As we have already seen, the antibacterial properties of honey help prevent wounds from becoming infected.

Case Thirty: Instructions concerning a sprain in a vertebra of the neck

If you examine a man having a sprain in the vertebra of his neck, you should tell him to look down at his two shoulders and breast. When he does so, you will be able to see how painful the sprain is and gauge its seriousness. You should cover the sprained area with fresh meat the first day. Afterward, you should treat it with honey every day until he recovers.

Again the Egyptian physician was right on, since honey's effectiveness as an anti-inflammatory has been confirmed by modern research.

The Smith Papyrus also explained how to treat burns with linen bandages soaked in honey, an area modern researchers have studied with positive results. Sycamore bark and anise seeds, when mixed with honey and water, produced an effective gargle to relieve sores and ulcers of the mouth. Malachite, a form of copper carbonate, was pulverized and mixed with honey to cure conjunctivitis, a common eye infection. The same mixture was used in eye makeup, a cosmetic the Egyptians applied lavishly, as evidenced in their stylized tomb paintings. Imagine—an eye makeup that doubled as a cure for eye infections.

The ingenious Egyptians also used honey as a laxative for chronic constipation and an anti-inflammatory for stiff joints. (Modern medical research confirms its efficacy in both of these conditions.) A popular birth control method called for a spermicide made of honey mixed with powdered crocodile feces; if the crocodiles weren't obliging, elephant dung could be substituted. Since modern science hasn't gone near this one, we can't comment on its efficacy, but in present-day Egypt, cotton soaked in a mixture of honey and lemon juice is still used by women as a prophylactic. The high osmotic pressure of the honey presumably kills the spermatozoa before they can complete their journey to conception.

One of the cornerstones of traditional Chinese medicine is *Shen Nong's Herbal Materia Medica,* a three-volume book compiled by Chinese scholars around the time of Christ to preserve the theories of Shen Nong, the much-revered herbologist who studied the efficacy of natural remedies about five thousand years ago. Shen Nong is said to have used himself as the guinea pig in his research—and when an herb proved toxic, he had to quickly discover its antidote.

According to Shen Nong, honey can "supplement insufficiency" in internal organs such as the heart, kidneys, and liver. Mixed with other medicines, honey "strengthens the will, firms the body, restores youth, and creates harmony." When added to the juice of fresh ginger and garlic, it helps relieve asthma. Mixed with ginger alone, it augments ginger's antibacterial properties.

During the Tang Dynasty (AD 618 to 907), beeswax was used to make pills easier to swallow, while bee stings were found effective in treating arthritis, a theory that is being tested in medical labs today. But it was honey, not beeswax or stings, that had the widest-ranging medicinal applications. In the treatment of smallpox, honey was rubbed all over the patient's body to stop the progression of the disease and help prevent scarring. Tang physicians also found honey effective in treating heart pain, sore muscles, and ulcers.

The Chinese emperors particularly valued royal jelly, believing it increased longevity and heightened sexual potency (the world's first Viagra?), a great boon when you have three or four wives and scores of concubines.

Even in modern China, most of the honey produced is used for medicinal purposes, such as the treatment of fluid deficiency, blood disorders, constipation, sore throat, and a general feeling of weakness. Many medicinal herbs are powdered, then mixed with honey as a binder to form pills. Taking a page from the Tang, the pills are then coated with beeswax. Some medicinal roots and leaves are stir-fried with honey to increase their efficacy.

During the Golden Age of Greece, Hippocrates, the father of modern medicine and author of *On Regimen in Acute Diseases,* written in 400 BC, contended that honey could be used to "soften hard ulcers of the lips and heal carbuncles and running sores." He proposed *oxymel,* a mixture of honey and wine vinegar, to "promote expectoration and freedom of breathing . . . it relieves the lungs and proves emollient to them and when it succeeds in producing these effects, it must do much good." But because vinegar can be acidic and irritate the intestines, "one should add to [the honey] merely as much vinegar as can just be perceived by the taste, for thus what is prejudicial in it will do the least harm and what is beneficial will do the more good." Hippocrates also recommends *hydromel,* a mix of honey and water, to moisten the mouth and quench the thirst. (It sounds as if Hippocrates invented the first boutique sports water or flavored power drink.)

In the fourth century BC, Democritus, known as the "laughing philosopher," credited honey for his long and healthy life—he laughed his way through 109 years.

In the Roman pharmacopoeia, honey was often prescribed alone or in combination with herbal ingredients. According to such scholarly sources as Pliny the Elder (AD 23–79), the Romans believed honey cured maladies of the throat and mouth as well as pleurisy, pneumonia, and even snakebite. Pliny himself suggested that honey mixed with the sap of aloe was an effective treatment for bruises, burns, and abrasions. Pliny's contemporary, Dioscorides, creator of *De Materia Medica,* the five-volume Roman equivalent of our *Physicians' Desk Reference,* was a great believer in honey as an agent of good health. Among other things, he recommended it for sunburn, ulcers, inflammations of the tonsils, and cough, and as an effective way to kill lice and nits. The centenarian Pollio Rumilius told Julius Caesar that his long life could be attributed to *interius melle, exterius olio*—honey on the inside, oil on the outside. Marcellus Empiricus, who lived in the Bordeaux region of France in the fifth century AD, wrote: *Honey, butter, and oil of roses, each in a like quantity, when warmed helps ear ache, dullness of sight, and white spots in the eyes.* Very prescient, in that modern researchers are now studying the efficacy of honey in the treatment of cataracts.

A couple of hundred years after the fall of the Roman Empire, the prophet Muhammad was quoted as saying: *Honey is a remedy for every illness of the body and the Quran is a remedy for all illnesses of the mind. Therefore I recommend to you both remedies, the Quran and honey.* Quite a testimonial.

The Quran itself explains: "There comes from within [the bee] a beverage of many colors, in which there is healing for men; most surely there is a sign in this for a people who reflect" (16:69). Also attributed to the Prophet is the belief that "there is a remedy in three things: a drink of honey, bleeding by a cupper, and cauterization by fire."

Many of the faithful religiously followed the Prophet's advice and regularly drank *sakanjubin,* a restorative tonic made from wine vinegar, herbs, and spices, and spiked with plenty of honey. Happily, the Quran does not forbid this type of wine-based drink, only *khamt* or true wine.

The Arabs of Muhammad's time also used honey to treat stomach ulcers, prevent cuts from becoming infected, and promote the healing of wounds without leaving scars.

Alexander's Final Journey

Not only did the ancients believe honey saved lives, but they may also have used it to embalm bodies from which life had departed. The ancient Aryans, Babylonians, Sumerians, Egyptians, and Cretans are all thought to have buried their great men in honey since honey not only conferred immortality, but was a great preservative thanks to its powerful antibacterial qualities. It has long been said that the corpse of Alexander the Great was embalmed in honey so it could make the long, arduous journey from the shores of the Euphrates River, where he died, to its final, faraway resting place.

Alexander the Great had conquered the known world by the age of thirty-three. His empire stretched from the Greek islands to the desolate mountains of the Hindu Kush in modern Afghanistan. In the sacred Egyptian oasis of Siwa, he had communed with the gods—and, in his own mind, had become a god himself. Yet gods are immortal, and Alexander the Great of Macedonia was not. Perhaps the ancient Greek and Egyptian deities thought him an interloper and decided to punish him for one of the worst sins of all, hubris. In any event, after a night of revelry in the ancient palace of Nebuchadnezzar on the banks of the Euphrates, Alexander was taken sick. It may have been a matter of too much unmixed wine or a dose of poison administered by a jealous rival. Or perhaps it was the work of the gods, reminding the Greeks who really rules. Whatever the cause, after ten days of feverish delirium, Alexander the Great, master of the world's mightiest realm, was no more.

According to a popular and enduring legend, the young king's ravaged body was embalmed in liquid honey, then placed in a massive sarcophagus of pure gold amid the mourning of his men and the keening of his women. In the first century AD, nearly four hundred years after Alexander's death, the Roman poet Statius refers to the embalming in one of his poems (*Silvae,* Book 2, #177). We can't say for sure where Statius got the story, but we can assume that it had been in circulation for generations.

Whether the legend is apocryphal or not, the fact is that honey would have been the perfect agent to prevent the decomposition of the corpse as it traveled across the deserts of the Middle East. Because of its osmotic properties, it literally sucks the life out of bacteria that cause dead flesh to decay.

Although the location of Alexander's tomb is uncertain—it could be Alexandria, the oasis at Siwa, or Macedonia itself—it is known that the golden sarcophagus was melted down for coinage by Cleopatra XIV and replaced by one of glass or alabaster. The great wealth interred with the king was undoubtedly looted in antiquity yet the search for his tomb by professional and amateur archaeologists continues to this day. If it is ever discovered, we may learn more about the role of honey in the preservation of his mortal remains.

The Use of Honey in Modern Medicine

Between 200 BC and AD 400, countless texts were written about the therapeutic advantages of honey, often recording discoveries made during the previous two thousand years. Yet despite its widespread use as a healing agent by the ancients, the curative properties of honey were largely forgotten by medical practitioners after the fall of the Roman Empire. We know of only one text, *The Leech Book of Bald,* written between AD 924 and 946 by an English monk named Bald, that recommends honey as a medicine, in this case as an eye salve and wound treatment (two uses being investigated by modern researchers). Virtually nothing was written about the medical attributes of honey during the Dark and Middle Ages, periods of intellectual stagnation in many disciplines.

With the coming of the Renaissance, however, there was a sporadic revival of interest in honey as medicine. An anonymous treatise written in 1446 describes a seven-step regimen for ulcer care, using a multitude of ingredients that include honey, beeswax, white wine, and red cabbage leaves. In 1623, an English minister, the Rev. Charles Butler, published *The Feminine Monarchie,* an important treatise on the lives and ways of honey bees. In the book, he promoted the use of honey as a disinfectant, cough medicine, eye salve, calming potion for gastric upsets, restorative drink, and laxative. Interestingly, all of these uses of honey are being studied today. In 1759, Dr. John Hill wrote a book whose title says it all: *The Virtues of Honey in Preventing Many of the Worst Disorders; Particularly the Gravel, Asthmas, Coughs, Hoarseness, and a Tough Morning Phlegm.* The opening paragraph of this worthy book laments the fact that honey had fallen out of favor in medical practice:

> *The slight regard at this time paid to the medical virtues of Honey is an instance of the neglect men shew to common objects, whatever their value . . . we seek from the remotest part of the world medicines of harsh and violent operation for our relief in several disorders under which we should never suffer if we would use what the bee collects for us at our doors.*

During the nineteenth century, as more and more pharmaceuticals became available, the use of honey in health care nearly ceased altogether. People wanted modern medicines developed by modern scientists in modern labs. Few believed that the old folk remedies of their grandparents and great-grandparents had any real curative powers. In Europe and America, honey was used merely as a sweetener to help bitter chemical medicines go down. It was also added to the herbal or mud poultices that were applied to wounds—and was surely the best ingredient in the mix, the only one with actual anti-infective efficacy.

In the twenty-first century, however, medicinal honey has begun to make a comeback as new clinical studies confirm many of the ancient healing powers of our favorite sweetener. We have summarized a few of these studies to give you an idea of the current status of honey as part of the modern medical armamentarium.

Honey and the Treatment of Cataracts

At least fifty million people suffer from the effects of cataracts, which are responsible for more than half of the world's blindness. Since pre-Columbian days, Mayan shamans have been using the honey produced by their beloved stingless bees to treat this debilitating eye disorder. By allowing the honey to ferment and then spiking it with plant alkaloids, they produce *balche,* a sacred drink offered to their gods during ceremonial rituals. The shamans simply mix the *balche* with water to produce therapeutic eyedrops.

Despite the apparent success of the Maya, modern scientists are just beginning to investigate the efficacy of honey as a topical anti-cataract agent. Though there is no solid data, there is some published research. Early clinical trials in Russia and Romania indicate that flavonoids (plant chemicals) found in honey may slow the progress of cataracts, though they do not seem to prevent them. And ongoing studies in New Zealand are exploring the potential of powerful antibacterial honeys such as manuka to reduce or delay the loss of vision due to cataract formation.

Honeyed Bandages for the Treatment of Severe Wounds and Burns

When it comes to antibacterial properties, clinical researchers have found that all honeys are not created equal. Most inhibit bacterial growth to some degree, due to their high osmolarity and the bacteria-hating hydrogen peroxide they produce. But only some honeys are potent enough to actually stop bacterial growth. One of the most powerful antibacterial honeys is manuka from New Zealand. Manuka contains phytochemicals derived from the nectar of certain plants that manuka-making bees visit during their foraging expeditions. This, added to the osmotic pressure and hydrogen peroxide, makes it one of the honeys smart bacteria don't want to encounter. Clinical trials in Australia and New Zealand have used manuka-impregnated dressings to produce remarkable results in infected wounds and ulcers that had been unresponsive to conventional antibiotics. One study reported that manuka dressings successfully resolved a surgical wound that had failed to heal over a period of thirty-six months, during which it was treated with both systemic and topical antibacterials as well as three surgical procedures. After being dressed with manuka bandages, however, the wound was completely healed in just one month.

Not only have manuka and other honeys been found to resolve persistent infections, they also clean and deodorize wounds and reduce pain, inflammation, and the level of exudate (the moisture that wounds emit, which can undermine the efficacy of topical treatments by diluting them).

Honey has been found particularly helpful in treating severe burns when it's applied to the dressing that covers the wound. (The Egyptians figured this one out about thirty-five hundred years ago.) Because they will not stick to the burn during scab formation, honey-impregnated dressings prevent the painful irritation caused by traditional dressings that adhere to the burned tissue. This, in turn, allows faster tissue regeneration and promotes skin growth.

A 1991 clinical study in New Zealand compared the efficacy of honey-impregnated bandages in healing burns to that of silver sulfadiazine, the standard medical treatment. After seven days, 91 percent of the burn patients treated with honey were free of infection, compared to only 7 percent of those treated with the silver compound. Within fifteen days, 87 percent of the patients treated with honey were completely healed, compared to only 10 percent of the sulfadiazine-treated patients.

Although the medical community was initially reluctant to accept the findings of these studies, the accumulation of positive results has begun to bring many physicians around. As more and more bacteria become resistant to conventional antibiotics, honey is gaining broader acceptance as a therapy for recalcitrant wounds and burns that haven't responded to pharmaceuticals.

This acceptance, along with the establishment of protocols outlining the proper usage of honey-impregnated dressings and specifying how often they should be changed, has spurred the commercial development of these products. In 2001, two honey-impregnated dressings came on the market for the treatment of severe burns and unresponsive wound infections. HoneySoft, manufactured by a Dutch company, is a patented plaster that contains a modern healing agent (ethylvinylacetate) and pure, high-grade honey free of pollutants. In one clinical trial, HoneySoft was used to dress sixteen traumatic wounds, twenty-three complicated surgical wounds, and twenty-one chronic, nonresponding wounds. All but two of the wounds had successfully healed in a mean time of three weeks (range one to twenty-eight weeks).

Mesitin, also produced by a Dutch company, is a dressing that contains a sterilized mixture of honey, lanolin, sunflower oil, and zinc oxide. Mesitin is very similar to Desitin, a product for the treatment of burns developed in Germany during the thirties but later abandoned for that purpose.

In Australia, New Zealand, the Netherlands, and to a lesser extent the United Kingdom, honey-impregnated dressings are now being used in many private and state hospitals. Though not yet available in the United States, which frequently lags behind other countries in adopting new therapies, they are being reviewed by the FDA and may soon win approval.

Honey and the Prevention of Fatal Hospital Infections

Thanks to its lethal effect on many dangerous pathogens, honey may soon play a role in the recovery of postoperative patients as well as patients who have received organ transplants and skin grafts. The most significant environmental factor that impedes the healing process in these patients is the presence of pseudomonas and staphylococcus, microorganisms that can cause fatal infections if left unchecked. Because honey has been proven effective against both pseudomonas and staphylococcus, there may be benefits to applying it on or beneath the sterile bandages used to dress surgical incisions and grafts.

Further Exploration of the Role of Honey in Modern Medicine

In the past two decades, there have been at least thirty-five clinical trials conducted in Australia, New Zealand, and the United Kingdom, involving more than six hundred patients, to study the efficacy of honey in promoting wound healing as well as in the treatment of skin conditions, stomach ulcers, and sore throats.

In 1988, a study of severe skin conditions such as bedsores and gangrene in patients who had not responded to conventional antibiotics found that those patients did respond to topical applications of raw honey. In most cases, the infections cleared within seven days.

A study conducted in New Zealand in 1994 found that applications of raw manuka honey stopped the growth of *Helicobacter pylori*, the bacterium responsible for stomach ulcers. More recent studies have found that manuka honey also halted the growth of *Streptococcus pyogenes*, a bacterium that causes sore throats in humans.

In Australia, honey is now officially registered as a "therapeutic good," the first time a food substance has gained such recognition. This means that its use as a medical/therapeutic agent is strictly monitored by the government. If a honey product is to be labeled as having antibacterial properties, its manufacturers must first provide clinical data supporting its efficacy.

Recently, 138 Australian honeys were tested for their bacteria-killing potency. Of these, 67 percent exhibited significant anti-bacterial activity. The most effective of these honeys were from the Australian jelly bush (*Leptospermum polygalifolium*). Sterilized jelly bush honey is already being marketed under the trade name Medihoney and has gained wide acceptance in hospitals in Queensland and New South Wales.

Modern medicine has just begun to tap the therapeutic potential of honey. One day soon its use in the fight against cataracts may be standard procedure. And its ability to promote wound healing and to kill bacteria already seems indisputable. Happily, scientists are continuing to investigate the curative powers of honey, including its anti-inflammatory properties, its ability to ease edema (swelling) and erythema (skin redness), its power as a promoter of collagen synthesis, tissue growth, and new skin formation, its potential as an anti-oxidant, its possible role in phagocytosis (when white blood cells attack invading pathogens), and much more.

It's interesting to note how many of the "newly discovered" medical properties of honey are not new at all. As we have seen, most of its therapeutic uses currently under investigation were known to physicians, shamans, and healers for many, many centuries. It seems that modern science has at last come to appreciate the value of a five-thousand-year-old folk remedy. Perhaps new doctors can indeed learn old tricks.

Afterword

A Letter to the Hive

It occurs to me, dear bees, that you may wonder why we chose to call this book about you and the honey you produce *Letters from the Hive* since you can't write and don't have easy access to stationery stores, post offices, and e-mails. Well, you may not be able to literally write and send letters, but you do have ways of communicating with us. As you go about your daily tasks, weaving through the air from flower to flower on balmy spring days, or swarming across the sky in a big black cloud shopping for a new place to live, or entering a nosedive on the way to delivering a sting to an inconsiderate human arm or neck, you are actually giving us lots of information about your world and the way you live in it—information that could help us live better in ours. Your sense of community service and devotion to the greater good, your efficiency and industriousness, your cozy, well-organized nests and attack-only-when-attacked policy, and above all the harmony with which you live within the rhythms of nature—these are qualities which would serve your human neighbors well if we could only manage to emulate them.

If we were to reply to your letters, care of the hive, there is a lot we could say. Some of the letters would be thank-you notes for all the honey you've produced and we've enjoyed over the millennia. Others would convey sincere apologies for our shameless plundering of your nests. It would be nice if we expressed an interest in experiencing your world. Perhaps you'd let us hitch rides on your fuzzy backs as you pay your amorous, pollen-gathering calls—those all-important visits that set bountiful harvests of fruit and vegetables on our tables and enrich our lives with seductive scents, brilliant colors, and unforgettable tastes, none of which would be possible without you. I imagine many of our letters would be full of irksome questions: what does the world look like to you, what do you make of the big, ungainly bipedal creatures who wreak such havoc on the planet, how does it feel to fly so high, what does honey taste like to the bee palate, do you ever balk at the dictates of your queen and long for a little individual self-expression, do you wonder what it would be like to drive a car, and what is it like to have sex on the wing?

I expect many of our letters to the hive would be unabashed love letters, for not only do we crave your honey, but we also love and admire you—so much, in fact, that we have painted your images on cave walls, enthroned you among our gods, set you up as political icons, invented rituals to honor and celebrate you, and applied the products of your industry to nearly every aspect of our lives, from meals to medicine, intoxicants to antioxidants, candles to lubricants. Who among us isn't enchanted to see you sipping nectar from among the petals of a flower? It's a very reassuring sight, nature at its best, the follies of our fellow humans set in perspective, the miraculous web of life reaffirmed.

Acknowledgments

It's not easy for a family to put up with the often quirky pursuits and eclectic habits of a field entomologist, but my family has always been wonderfully understanding. Jane, my late mother, tolerated my entomo- and dinocentric childhood, providing endless empty jam jars to house the latest bug du jour and sewing bridal veil material onto clothes hanger hoops to make my first insect nets. My late father, Stanley, accompanied me on camping trips around California and offered encouragement and support early on in my academic career. My sister, Lisa, put up with older-brother harassment and frequent practical jokes while never losing her sense of humor. My daughters, Marlyse and Melissa, accompanied me on many a wondrous field trip in search of native bees and other Sonoran desert animals and plants. I take great pleasure in dedicating this book to them.

A book on bees and honey would not have been possible without the life-enriching associations I have had with my scientific colleagues: John Alcock, John Ascher, Robert Brooks, James Cane, Gabriella Chavarria, Bryan Danforth, Terry Griswold, Bernd Heinrich, David Inouye, Eugene Jones, Norbert Kauffeld, Carol Kearns, Peter Kevan, Claire Kremen, Gretchen LeBuhn, Ronald McGinley, Charles Michener, Robert Minckley, Gary Paul Nabhan, Jack Neff, Chris O'Toole, Laurence Packer, Anthony Raw, David Roubik, T'ai Roulston, Jerome Rozen, Justin Schmidt, Charles Shipman, Avi Shmida, Hayward Spangler, Evan Sugden, Vincent Tepedino, Robbin Thorp, Philip Torchio, Rogel Villanueva, William Wcislo, Edward Wilson, and Doug Yanega. Thank you one and all for much good fun and camaraderie.

For teaching me the ways of the honey bee, I gratefully acknowledge and thank my mentors in the biology and behavior of the genus *Apis:* Paul Cooper, Makhdzir Bin Mardan, Robert Schmalzel, Thomas Seeley, Hayward Spangler, Steven Thoenes, and Adrian Wenner to name but a few. Thanks also to my favorite Arizona beekeeper, Leonard Hines, for his part in making the television documentary *Pollinators in Peril* and for his good-natured ways, hearty radio-announcer-like voice, and beckoning smile. I thank Peter and Becky Fonda for bringing blue orchard bees to their organic apple orchard in Montana, for making the film *Ulee's Gold,* and for becoming outspoken advocates of pollinator conservation.

Scott Hoffman Black, Gretchen Daily, Sam Droege, Gretchen LeBuhn, David Hancocks, Claire Kremen, Robert Pyle, Matthew Shepherd, Chip Taylor, Steve Walker, Neal Williams, Mace Vaughan, and others have boldly advanced pollinator conservation, research, and educational outreach programs. I especially applaud the Xerces Society for all that they do to foster invertebrate conservation around the world, and I am proud to serve as one of their counselors. Recently, they and others founded the North American Pollinator Protection Campaign to stop or slow pollinator declines. A special thanks to Paul Growald, Gabriella Chavarria, and Laurie Adams for keeping the NAPPC vision bright and strong.

For funding many worthwhile projects, including native bee censuses and pollinator conservation, I gratefully acknowledge the staffs of the Wallace Family Foundations, the Geraldine R. Dodge Foundation, and Island Press. Thanks to Robert E. Turner and his sons, Rhett and Beau, for their efforts on behalf of pollinators, and to Mike Philips of the Turner Endangered Species Fund and to the staff and board of the Turner Foundation in Atlanta for their good work in sustaining pollinators, thereby ensuring a healthy and stable food, fiber, and fuel supply for the people of the world. Special thanks are extended to Melanie Adcock of the CS Fund and Diana Cohn of the Solidago Foundation for funding our recent research and educational efforts in the state of Quintana Roo, Mexico.

I sincerely thank many dear friends, some who have moved away but most of whom are still bona fide Tucson desert rats, for spirited discussions about bees and honey, and for sharing many fine adventures: Beth Armbrust, Kim Buck, Patricia Cowan, Alison Deming, Arthur Donovan, Mark Dimmitt, Richard Felger, Anne Gondor, Steve Hopp, Albert Jackman, Barbara Kingsolver, Donna McAlister, Margaret McIntosh, Sarah Richardson, Hayward Spangler, Cristina Trimble, and Michael Wilson. Special warmth and a jar of honey go out to Diana Cohn and Joshua Mailman, fine friends and co-founders with me of The Bee Works.

I, along with the publisher, thank Dr. Richard Jones and Dr. Eva Crane of the International Bee Research Association in Cardiff, Wales, for permission to reprint antiquarian art.

And thanks to the National Honey Board for permission to reprint their recipes and for all the good work they do on behalf of beekeepers.

A very special thanks to Richard Jones, executive director of the International Bee Research Association, and its former director, Eva Crane, for permission to reprint illustrations that are under their care and direction.

I would also like to thank the numerous librarians, especially at the University of Arizona's main and science libraries, who helped me chase down many obscure facts and figures. Any errors or misstatements that remain are mine alone. I thank Christine Cairns, Harriet J. de Jong, and Julio Lopez for their inspired writings about the Mayan agricultural system, meliponiculture, and endangered traditions. And I thank Troy Sagrillo for providing me with several unpublished photographs of Egyptian tomb and temple art and for guiding me through the often convoluted Egyptian archaeological evidence as it relates to bees, honey, and beekeeping.

A special thanks goes to my agent and manuka honey devotee, Judith Riven of New York, who was dogged in her insistence, over a period of more years than I care to admit, that this book come into being. Thank you, Judith! To my co-author, Banning Repplier, are extended the warmest of thanks and gratitude for climbing aboard the bee/honey wagon and following the bloom through endless rewrites while never complaining. Thank you, Banning. To my editor, Beth Rashbaum of Random House, I extend my sincerest thanks and gratitude for her unwavering support, fine editorial suggestions, and wordsmithing, and for shepherding this book from the proposal stage to its final publication. To Paul Mirocha, my longtime Tucson friend and fellow traveler to the Malaysian rainforests, we extend our thanks for his wonderfully sticky cover art. Paul is the king bee of pollinator art, and a fine mandolin picker.

Lastly, I thank the various honey bee species of the world, as well as the more than twenty thousand other bee species, for providing me not just with an unusual career but with endless opportunities to be fascinated, humbled, and thoroughly amazed by the spectacle of life on earth.

Appendix 1: Glossary

Africanized honey bee (*Apis mellifera adansonii*). Originally from Central and South Africa, Africanized honey bees were brought to Brazil in 1957 and then escaped, migrating north and eventually spreading throughout the central and southern regions of the United States. They defend their nests vigorously and swarm frequently at any time of the year.

American foulbrood. A bacterial disease, caused by *Paenibacillus larvae,* that attacks honey bees. It produces a foul odor in the bee larvae it kills, hence its name. European foulbrood is a related disease.

Apiarist. A person who manages honey bees in man-made hives. Also known as a beekeeper.

Apiary. The place where beekeepers keep several hives containing honey bee colonies.

Bee bread. Pollen mixed with nectar and saliva and stored in special areas of the comb. Microbes help to ferment the bee bread into a stable food for the bees.

Beeline. The more or less direct route flown by a worker bee from its hive to a nectar source and then back home again. Bees actually drift and curve in their flight paths, especially during windy periods or when they have to avoid obstacles.

Bee space. The ideal amount of space (1/4 to 5/16 inch) between the combs in a natural or man-made hive. Discovered by Lorenzo Langstroth of Pennsylvania during the Civil War era.

Beeswax. A complex substance, resembling fat, secreted from four paired glands on the abdomens of young worker bees and used in the construction of honey and brood combs.

Bobcat. Not a carnivorous mammal but a small forklift used by modern beekeepers to load and unload hives from their trucks.

Brood. The eggs, larvae, and pupae in a honey bee colony are collectively known as the brood.

Brood comb. A comb made up of waxen cells containing eggs, larvae, pupae, and emerging young adult bees.

Brood nest. A basketball-sized area of brood comb in the lowermost supers of a Langstroth hive. Sometimes called the brood sphere. Here is where you find the youngest bees and the egg-laying queen.

Cell. The individual unit that makes up a comb. Each hexagonal cell is formed from beeswax and has a tapering triangular base. Cells are joined to the sides of neighboring cells and to the bottom of cells on the other side of the comb. Almost everything that happens within the colony takes place in its myriad cells. Talk about cubicles in the workplace!

Chunk honey. Unprocessed, unfiltered honey just as the bees made it, still in the capped honeycomb. The pieces of comb are usually sold in a glass jar filled with liquid honey, or sometimes in a small wooden box.

Cluster. A tightly knit mass of bees, formed when they are making wax or to keep warm during the cold winter months.

Colony. The honey bee community, made up of castes (workers, drones, and a queen). Sometimes called a superorganism because it functions as if it were a single unit. A colony may have as many as sixty thousand individuals during the peak season. The place where the colony lives is called the hive or nest.

Creamed, spun, or whipped honey. A thick form of honey, as opposed to the liquid most of us are familiar with. A special commercial process (the Dyce process) seeds honey with microscopic crystals, transforming it into a granulated spread.

Crystallization. The formation of tiny crystals in honey as it ages, eventually turning it into a solidified mass. Some honeys, however, such as tupelo, rarely crystallize.

Cyser. Honey wine made from fermented honey and apple juice or cider.

Drone. Male honey bees whose sole purpose in life is to provide the queen with sperm. They don't forage for nectar or pollen and are fed by their sisters.

Drum. A fifty-five-gallon steel barrel used by beekeepers to transport their honey crop to cooperatives and other honey packers.

Dyce process. The process of controlled crystallization used to produce creamed or spun honey, discovered and patented by Professor E. J. Dyce of Cornell University in 1935.

Excluder. A wire frame placed between the lower brood supers and the upper honey supers to prevent the queen from laying eggs in the honey frames.

Extracted honey. Honey that has been removed from the frame with a centrifugal extractor.

Extractor. A hand-powered or electrical device that spins honey frames at high speeds to extract the honey. Centrifugal force sucks the honey out of the uncapped combs.

Foundation. Hexagonally imprinted, man-made sheets of beeswax, provided by beekeepers to serve as a foundation on which the bees can build their combs.

Frame. A rectangular wooden frame (measuring 17 5/8 by 9 1/8 inches) used to hold prefabricated sheets of hexagonally imprinted beeswax and/or plastic comb foundation.

Granulation. A natural process that occurs as honey ages. Various impurities, such as yeast cells and pollen grains, create nuclei around which sugar crystals form. Honey can be made liquid again by gently heating it in a double boiler or on a low setting in a microwave oven.

Haplodiploidy. An unusual method of sex determination in bees, ants, and wasps. The queen can control whether the eggs she lays are fertilized or not. Female worker bees hatch from fertilized eggs and are diploid, having two sets of chromosomes. Male drones hatch from unfertilized eggs and are haploid, with only one set of chromosomes.

Hippocras. An ancient form of mead, made with fermented honey, grape juice, and various spices such as cinnamon, ginger, and nutmeg. Once used as a medicine.

Hive. The hollow tree, Langstroth box, or natural cavity where a honey bee colony lives. Also referred to as the nest.

Honey bee. A social, honey-making bee, one of approximately eleven species in the genus *Apis,* found mostly in the tropical regions of Asia.

Honeycomb. Comb found in the upper supers of a Langstroth hive and used for storing honey. When a queen excluder is placed between the brood nest and the upper supers, the queen cannot lay eggs in honeycombs and the wax remains white in color, unlike the dark brown brood combs.

Honey sac. A balloonlike sac in a bee's foregut. It acts as a temporary storage area for nectar that is to be carried back to the nest.

Invertase. An enzyme produced from yeasts by the hypopharyngeal glands in the heads of worker bees. It is added to honey during the ripening process, where it splits sucrose into its component parts, fructose and glucose.

Langstroth hive. A type of wooden hive or nest box with movable frames and the right amount of bee space. Invented by Lorenzo Langstroth in the 1860s.

Larva. The bee grub, which goes through five molts before entering the pupal stage.

Mead. Honey, water, and other ingredients such as herbs and spices that have been fermented into honey wine. Probably the world's oldest alcoholic beverage.

Melomels. Honey wines made with fruit or fruit juices in addition to honey.

Migratory beekeepers. Beekeepers who transport their hives across the country in flatbed and eighteen-wheeler trucks. They follow the bloom to produce specialty honey crops or provide lucrative pollination services.

Nectar. The sugary, often complex watery mixture produced by special secretory glands in flowers and extrafloral nectaries. A high-energy fuel exchanged for pollination services.

Nosema. A bee disease caused by a protozoan, *Nosema apis,* which can be a problem for beekeepers in some areas of the United States.

Nurse bee. A young honey bee worker six to twelve days old with enlarged hypopharyngeal glands that secrete royal jelly. Nurse bees feed the royal jelly to developing larvae.

Oxymel. Mead made with honey and wine vinegar.

Package bees. Beekeepers wishing to start a new colony typically purchase a queen and a two- or three-pound "package" of workers and drones from a queen breeder and package bee supplier. If you don't catch and hive a swarm, this is the easiest and fastest way to start a new colony.

Pheromone. Chemicals used by many insects to mark food sources, warn their neighbors of danger, or secure a mate. Common honey bee pheromones include Nasonoff pheromones, used to scent-mark the hive; alarm sting pheromones, the bananalike scent that causes bees to become alarmed and sting intruders; and the famous queen pheromone, which causes workers to feed and care for her and also attracts drones to mating areas.

Pollen. Microscopic, tough-shelled grains that contain the male sex cells of flowering plants (angiosperms). For fertilization to occur, the grains must be transferred by pollinators from one flower to another. The sex cells travel down pollen tubes in the floral style and fertilize the ovules. Seeds develop within and the fruit forms. Pollen grains are also the indispensable protein- and lipid-rich food of bees.

Pollination. The transfer of pollen grains from flower to flower and plant to plant by wind, water, and bees and other pollinating insects and animals. Fertilization and seed production follow.

Pupa. The so-called resting period between the larval and adult stages in bees and other insects. During this period, dramatic biochemical changes transform the larval tissues into those of an adult.

Pyment. Mead made with both honey and grape juice.

Queen bee. The queen of the honey bee nest. Usually there is just one queen in a mature bee colony. She is the mother of them all, laying fifteen hundred eggs a day during her lifespan of two to three years.

Ripening. The process by which bees transform dilute floral nectar into thick golden honey. The bees add enzymes (invertase, diastase, and glucose oxidase) to the nectar and evaporate the excess water by fanning their wings over the mixture and by sucking it into their mouths and crops and regurgitating it multiple times. Ripened honey is stored in the capped cells of the honeycombs.

Royal jelly. A secretion produced by the hypopharyngeal glands in the heads of young worker bees. Larvae destined to become queen bees are fed entirely on royal jelly, while worker and drone larvae are fed on royal jelly for a few days, then switched to pollen and honey.

Scout bee. Foragers who find sources of nectar and pollen. Successful scouts return to the nest and inform the colony where the food sources are by performing waggle dances and/or through the floral odors clinging to the hairs on their bodies.

Smoke. For millennia, honey hunters and beekeepers have used smoke to calm the bees as they extract the honey from their nests. When the bees smell smoke, they think the nest is burning and gorge on honey as they prepare to flee. The large amount of ingested honey causes bloating, which renders them unable to sting the invaders.

Split or divide. Beekeepers create new colonies by dividing the adults, brood, and honeycombs of an overpopulated colony and housing them in separate Langstroth hives. The beekeeper either supplies a packaged queen to the daughter colony or allows the bees to be queenless while they raise a new queen from an egg.

Super. A wooden box in a Langstroth hive that contains frames on which the bees construct their combs.

Swarm. A mass of flying bees consisting of a queen and about half of her daughters, the worker bees. The swarm emerges from an overpopulated nest and flies off to a new hive site.

Syrup. A cane sugar, beet sugar, or corn syrup solution fed to honey bee colonies to compensate for nectar shortages, or to help build up colony populations in the early spring.

Varroa mite. A giant parasitic mite *(Varroa jacobsoni)*, introduced into the United States from Asia, which destroys colonies by feeding on larvae and pupae in the brood combs.

Volatiles. The essential oils and aromatic chemicals that give flowers, nectar, and honey their characteristic scents. Overheating honey can drive off these appealing fragrances.

Waggle dance. A dance performed in the hive by returning scout bees, thought to direct other bees to the floral supermarkets they have located. Though noted by Aristotle and other early bee watchers, waggle dancing was not scientifically investigated until the 1930s by the late Austrian Nobel laureate Karl von Frisch.

Water white. The official color designation of the mild commercial honeys that most Americans seem to prefer. (Give me a dark, rich honey any day.)

Worker bee. One of the queen's thousands of sterile daughters. Workers typically toil inside the nest as nurses, wax makers, and honey ripeners, then outside it as foragers, living for only four to six weeks during the busy spring and summer months.

Appendix 2: Bees of the World: A Remarkable Beestiary

Bees are distinguished from wasps and ants, their relatives in the order Hymenoptera, by certain physical and behavioral characteristics. They are covered with hairs that are plumose (branched like feathers), all the better to grab on to pollen. They also have a constricted waist, which gives them flexibility when turning in their narrow burrows. It is believed that over one hundred million years ago, during the rise of the angiosperms, or flowering plants, the first bees evolved from predatory hunting wasps. Although some of the world's bees are partial meat eaters, the vast majority is content to be vegan, dining on sweet hoards of nectar, pollen, and honey.

Most bees are solitary creatures. They do not live in complex societies in large nests and they do not store honey. Instead, they do their own thing. Each female is totally self-reliant, excavating her own nest and foraging for her own nectar and pollen, which she eats on the spot or brings back to her nest for her developing larvae to consume right away. European honey bees (*Apis mellifera*) and a few others such as bumblebees (genus *Bombus*) and the stingless bees of the Maya (*Melipona* and *Trigona*) are the exceptions. They store the honey they produce. Modern beekeepers aid this natural tendency by giving honey bees more room in their hives than they actually need, encouraging them to produce the surplus that eventually finds its way to our tables.

The scientists who study bees are properly referred to as melittologists (apiculturists are experts who study only honey bees). These experts are confronted with a gargantuan task. *The Bees of the World,* a 913-page book published in 2000 by Charles Michener, recognizes 7 families and 425 genera of bees worldwide. Within the genera, almost 20,000 bee species have been described so far. Estimates of how many species are really out there, including those yet to be discovered and formally named, range from 25,000 to 40,000. There are about 4,000 bee

True Honey Bees

Genus *Apis,* divided into three subgenera: *Micrapis, Megapis,* and *Apis*

Subgenus Micrapis: *Small or Dwarf Honey Bees*

Red dwarf honey bee *(Apis florea)*
  Habitat: Arabian Peninsula to Sri Lanka and Southeast Asia
Black dwarf honey bee *(Apis andreniformis)*
  Habitat: Java to South China and Bangladesh

Subgenus Megapis: *Giant Honey Bees*

Giant Himalayan honey bee *(Apis laboriosa)*
  Habitat: Laos and South China to Nepal
Giant Southeast Asian honey bee *(Apis dorsata)*
  Habitat: Rainforests of Southeast Asia
Giant Sulawesi honey bee *(Apis binghami)*
  Habitat: Sulawesi archipelago
Giant Philippine honey bee *(Apis breviligula)*
  Habitat: Northern Philippine Islands

Subgenus Apis: *The European Honey Bee and Allies*

Western or European honey bee *(Apis mellifera)*
  Habitat: Introduced by beekeepers to most continents
Eastern honey bee *(Apis cerana)*
  Habitat: Oman to Timor, Japan, East Russia, the Philippines, Sulawesi, Borneo, and Taiwan
Sundaland honey bee *(Apis koschevnikovi)*
  Habitat: Southern Thailand to Sumatra
Sulawesi honey bee *(Apis nigrocincta)*
  Habitat: Sulawesi archipelago
Malaysian mountain honey bee *(Apis nuluensis)*
  Habitat: Mt. Kinabalu, Borneo

species in the United States, 7,000 in South America, roughly 4,000 in Africa, and 3,000 in Australia.

Given these numbers, it is not surprising that the world is literally full of bees—bees of all imaginable sizes, shapes, colors, habits, and habitats. From sea level to altitudes of over fourteen thousand feet, bees have invaded and colonized every part of the planet that offers flowering plants and places to nest. Only Antarctica and the northernmost islands of the Canadian tundra are without bees.

True honey bees (genus Apis)

The first known honey bees date back forty million years to the Eocene Age. Specimens of later honey bee species in this genus have been found preserved in amber jewelry in Germany, Poland, and Scandinavia. Despite this evidence of northern exposure, however, most honey bees seem to have evolved in tropical lowland forests, where they are still abundant today. The familiar European honey bee (*Apis mellifera*) is believed to have originated in the African tropics, then migrated westward into Asia and northward to colder European climates. Until modern times, *Apis mellifera* was not found anywhere in the Western Hemisphere, Australia, or the remote Pacific islands, but thanks to enterprising European settlers who imported them for commercial purposes, they are now global citizens. In the United States, the honey bees that buzz through our gardens are the result of both planned and unplanned introductions to our shores. Many are the descendants of forebears who were deliberately brought to Mexico by Spanish beekeepers; others came aboard English ships laden with colonists and their homesteading supplies, bound for Jamestown and Plymouth Rock.

Depending on which bee taxonomist you talk to, there are anywhere from seven to nine or even eleven species of true honey bee, all members of the genus *Apis*. (In this book, I have considered the genus *Apis* to comprise eleven living species.) The first scientist to recognize, classify, and name honey bees was the Swedish naturalist Carl von Linné, better known as Linnaeus. He described the genus *Apis* in the year 1758. The actual specimen described by Linnaeus 246 years ago can be seen today in the British Museum of Natural History, skewered on a pin above a faded label written in Linnaeus's own hand. Having miraculously survived fires, floods, world wars, carpet beetles, mold, and human negligence, it remains the gold standard—the prototypical honey bee specimen to which all others of its kind are compared.

The eleven species within the honey bee genus belong to one of three subgroups, giant honey bees (subgenus *Megapis*), dwarf honey bees (subgenus *Micrapis*), and European and related honey bees (subgenus *Apis*). Most of the bees in these subgroups originated in the tropical regions of Asia. Only *Apis mellifera* and *Apis cerana*, have been "domesticated" and kept in man-made hives, although all have been robbed of their honey by human and animal predators.

There are two major species of giant honey bee in the subgenus *Megapis*: *Apis laboriosa* and *Apis dorsata*. *Apis laboriosa*, the world's largest honey bee, prefers to nest on remote cliffsides in the Himalayas of Nepal. The worker's body ranges in length from 0.67 to 0.75 inch and their colonies number from twenty thousand to sixty thousand, all living in harmony on massive, parabola-shaped combs. *Apis dorsata* nest high in the trees of tropical rainforests such as those in Malaysia, Thailand, Vietnam, and Borneo, although some have made the move to man-made structures—such as tall buildings and water towers—in those regions.

Colonies of *Apis dorsata* must always be approached with extreme caution. The movement of people walking far below their arboreal nests can trigger a massive and potentially life-threatening attack by their vigilant guard bees. *Apis dorsata* form living blankets covering both sides of combs that can be up to five feet across. Honey is stored in the wider, uppermost portions of the nest, where it is attached to the underside of massive branches. Favorite trees, such as the towering tualang tree of Malaysia, are often home to not one but dozens of colonies. Individual tualang trees are known to have housed as many as 120 distinct nests.

Although the extreme height of tualang trees would seem to offer all the protection the giant bees could need, the lowland Asian rainforests are also home to the giant sloth or honey bear. These greedy creatures come well equipped for nest robbing, with strong claws that facilitate tree climbing and a foot-long tongue, ideal for lapping up quantities of sweet honey and protein-rich brood. Another worry for the bees are honey buzzards, which circle above nest-bearing trees, waiting for an opportunity to plunder the giant combs. And then there are the human honey hunters, such as Pak Teh and his clan.

Except for some recent experimentation with "pole hives" in the forests of Vietnam, *Apis dorsata* are not managed. They are simply pillaged, often on a sustainable annual basis, by honey hunters in several countries.

A still-mysterious aspect of the giant honey bee story is their tendency to migrate great distances, probably following the bloom in search of the best nectar flows. In the Pedu Lake region of Malaysia, the first *Apis dorsata* bees return to their favorite tualang trees in November or December from unknown parts. They immediately begin to produce wax to build the combs where they will store their honey and rear their brood. In late February or March, the colonies begin to depart, one by one, to begin their mysterious migrations. The spectacle and roar of these huge colonies leaving their giant combs is a sight one is never likely to forget. Sailors and ship captains have reported seeing migratory swarms crossing the Straits of Malacca from the Malaysian mainland to the island of Sumatra.

There are two species of dwarf honey bee (subgenus *Micrapis*), one of which, *Apis florea,* is only 0.35 inch long—the length of a European honey bee's abdomen. Unlike the European honey bee *(Apis mellifera),* these bees don't make multiple combs or nest in dark cavities. Instead, they build a single comb, usually attached to the small branches of living plants. As with their giant *Apis* brethren, the surfaces of the fragile combs are protected from the rain, sun, and wind by a living blanket of interlocked bees several layers deep. *Apis florea* are rarely seen by people, for they like to hide their colonies, at most six thousand strong, in dense foliage.

Apis florea are relatively gentle bees, small in size with a correspondingly small stinger, which they often have difficulty implanting in human skin. The worker bees moonlight as security guards, ready to fend off attackers such as the rapacious weaver ant, a constant threat to all the bees in these Asian forests. As a further deterrent, they collect sticky plant resins and apply them to the branch supports to prevent the bothersome ants from infiltrating their colonies. But when under severe attack by a mammal or bird, these small, gentle bees abscond, abandoning their nests and brood to their fate.

Among the honey bee species, *Apis florea* store the smallest amount of honey (only about one quart per year), yet even this diminutive reserve has been plundered for millennia by humans and other honey-loving animals. In open-air food markets in Bangkok, you can often find *Apis florea* nests for sale, still attached to their branch supports and arranged in an attractive display resembling a tiered tree. These nests, succulent with honey, are gathered from the wild, since *Apis florea* are not managed by humans. They provide a special treat for many Thai families and are greatly prized.

Non-Honey Bee Bees

From the World's Smallest Bee to the World's Largest

The world's smallest bee, *Perdita minima,* can be found right in my front yard in Tucson, Arizona. Its body is 0.078 inch long, small enough to crawl right through the fine mesh of most insect nets, enabling it to avoid capture by curious entomologists. I usually spot this bee by the faint shadows it casts on roadsides and sidewalks. It pollinates the tiny white flowers of the native mat euphorbias. These weedy plants poke up through cracks in sidewalks around town and out in the desert proper. A stingless bee from the Amazon may be even slightly smaller than *P. minima.* One can imagine these Lilliputian bees struggling to secure a single pollen grain to bring home to feed the kids.

The world's largest bee, *Chalicodoma pluto,* is 1.6 inches long and shares its nest with arboreal termites in the Moluccas, the famed Spice Islands of the Orient. The first specimen of this giant, collected in the Moluccan rainforest in 1859 by Alfred Russel Wallace, the co-discoverer of evolution by natural selection, is housed in the British Museum of Natural History in London. This elusive bee was later lost to science until it was rediscovered in the 1970s by Adam Messer, then a graduate student at the University of Georgia. Little is known about *Chalicodoma pluto* beyond the fact that it is a member of the leaf-cutter family of bees, who cut neat snippets of leaves with their mandibles to use in the construction of their nests. (Other members of this family may have visited the rosebushes in your garden, leaving visible traces of their handiwork behind.)

More common are the gentle giants known as carpenter bees (genus *Xylocopa*). Along with bumblebee queens and some orchid bees, these are the largest bees in North America. They excavate extensive galleries in dead wood with their powerful mandibles and return to them year after year.

Cuckoo Bees, or Bees in Wolves' Clothing

These bees are very colorful and sleek, with little or no pelage of fine hairs. That's because, thanks to their nefarious lifestyle, they don't need fuzz to help them transport pollen. Instead of making an honest living, building their own nests and collecting their own pollen and nectar, they prey on others. Scientists call them cleptoparasites; we call them cuckoo bees, for like cuckoos and cowbirds in the bird world, they sneak their eggs into someone else's nest, usually when the owner is out shopping at the floral grocery store. The parasite egg hatches first and the young cuckoo bee larva sidles up to the host's eggs, kills them, and eats them. Next it gorges on all the pollen and nectar that the host mother had gathered for her own young. The cuckoo bees then emerge, mate, and seek out new host nests to begin the cycle all over again. It may not seem fair, but evolution has helped make the cuckoo bee a successful creature.

Polyester Bees

The Colletidae family of bees line their underground tunnels and cells with a clear, durable material not unlike cellophane. This ensures that the level of humidity in the nursery will be just right for the eggs and developing larvae. Bees are master chemists, and colletids are no exception. Some of them produce chemical compounds known as macrocyclic lactones. You might know these substances by their other name, polyester. Colletid bees may not be wearing retro leisure suits of the 1970s, but they are polyester-making bees nonetheless.

Mining Bees

Each spring, university entomology departments are besieged with phone calls from homeowners wanting to know who is responsible for the little mounds of soil mysteriously appearing on their lawns. The culprits, it turns out, are mining bees, which dig shafts below the ground in which to build their nests. The architecture of these nests can be quite elaborate, with just one cell or many dozens of them.

Mason Bees

Not with trowels but with mandibles and specialized tools on their legs, some bees are adept at collecting and shaping mud, resins, masticated leaves, and pebbles into safe, waterproof nurseries for their developing larvae. These bees don't dig their own nests; they simply redecorate abandoned beetle burrows with their own cell partitions and end plugs. The blue orchard bee (*Osmia lignaria*) is one such mason, common to the Pacific Northwest of the United States. It is also a champion pollinator of fruit trees, especially apple, plum, and sweet cherry—more efficient, in fact, than the industrious honey bee.

Sweat Bees

Ever been gardening and had a small black or metallic bee land on your forearm and lick your perspiration? If so, you've already encountered a sweat bee of the Halictidae family. This family includes solitary species as well as primitively social kinds that have worker castes and queens. These largely ground-nesting bees pollinate numerous kinds of flowers.

Other Honey Makers: Ants and Wasps with a Sweet Tooth

Some ant and wasp species take their cue from the bees, making and storing honey in their nests. The so-called honeypot ants (*Myrmecocystus* spp.) of the American southwestern desert collect nectar and honeydew from scale insects, then stockpile the liquid honey in a very unusual way. Certain citizens of the colony, called repletes, actually act as living honey jars. They greet the returning foragers and tap out a message in ant Morse code, which causes the foragers to transfer the honey to them. The abdomens of the repletes swell with the transferred honey until they are grotesquely round and distended. These living tankards hang patiently from the roof of the dark, underground nest until they are tapped into service and regurgitate the honey to their sister ants.

I remember an interesting encounter with honeypot ants near the Arizona–New Mexico border. I had been staying at the nearby Southwestern Research Station, operated by the American Museum of Natural History, while studying buzz-pollinating bees who visit deadly-nightshade blossoms. One morning, on my way to a nightshade patch, I stopped alongside an excavation in progress and met Harvard myrmecologist Gary Alpert.

Using a John Deere backhoe belonging to a nearby cattle rancher, Gary was busily excavating a honeypot nest—a Herculean task, since the ants often burrow fifteen feet or more into the rocky desert soil. He eventually got to the bottom of the nest and broke into galleries containing dozens of ant repletes. When Gary popped one of the ants into his mouth, I followed suit. After all, Aborigines in Australia have dined for millennia on sweet repletes. Who was I to break with such a venerable tradition? I held the squirming ant by its head and legs, popped its ready-to-burst abdomen with my teeth, and felt the acidic honey squirt out onto my tongue. Acidic (perhaps due to formic acid), yes, but delicious. As I tossed the emptied ant husk away and grabbed another, I wondered what flowers or scale insects the ants had visited and how far they had traveled in order to collect the honey for our impromptu feast.

A few tropical wasps (genus *Brachygastra*) are also in the honey business. They live in subtropical Texas, Mexico, and Central America in delicate but durable paper nests. Their open-faced, upside-down paper combs, enclosed in a protective outer sheath, are filled with dark honey, almost molasseslike in taste. One species, *Brachygastra mellifica,* from the dry, deciduous forests of Guanacaste province in Costa Rica, lives in nests housing the tens of thousands of individuals. They are known to defend their honey-laden nests fiercely when humans get too close. Interestingly, like honey bees and certain harvester ants, they are part of a select group of hymenopterans that have barbed stingers. Having a barbed stinger, or aculeus, has its drawbacks, since the stinging individual loses its life along with its stinger. (The females—the guys don't sting—of most bee, wasp, and ant species have unbarbed, smooth stingers, similar to hypodermic needles, that can be jabbed again and again to deliver their venom.)

Some tribes in northwest Mexico seek out *Brachygastra* bees and make off with their honeycombs. They chew the combs, spitting out the paper bits, then go back for more—a real treat in a region offering little in the way of natural sweets.

Appendix 3: Other Products of the Hive

When anyone mentions bees, our thoughts naturally turn to honey, but there are many other important substances collected or produced by honey bees and stored away in their nests. This appendix is about those products, which we'll examine in some detail.

Beeswax

Beeswax is the concrete, steel, and glass of the nest. With it, the clever bees construct walls, ceilings, honey and pollen pantries, nurseries for their brood, and even waggle dance floors.

In the first scientific study of the efficiency of beeswax production, fellow researcher Justin Schmidt and I found that it is biochemically very expensive to produce. We fed a mixture of 50 percent sugar and 50 percent water to caged honey bees of wax-producing age. Suspended in festoonlike chains of interlocking legs, the young bees guzzled the sugary solution and began slowly secreting wax from four paired glands on their bellies. We dutifully measured their wax output by weighing the brilliant white combs they had made during the course of our experiment. Based on these measurements, we determined that bees must ingest and metabolize twenty pounds of honey to make one pound of wax, turning simple sugars into the complex lipids. At this rate, it might take eighty thousand or more young worker bees to produce sixty pounds of fresh wax.

If you examine a worker bee with a microscope or hand lens, you'll see four paired wax glands on the underside of her abdomen. On the outer surface of the glands are flat, shiny oval areas known as wax mirrors. The wax is secreted as a clear liquid that floods the wax mirrors, where it quickly hardens into scales and turns a translucent white. (It takes 910,000 scales to make 2.2 pounds of beeswax.) The wax-making bee or one of her sisters eagerly grabs the wax scales and takes them to the comb construction site. To make the walls of the cells and the caps that seal them, the bees manipulate the scales with their mandibles and add enzymes and saliva to the mix. The now pliable wax is fashioned, without a blueprint, into the individual cells, which are joined side to side and end to end to form the double-sided combs. Each waxen cell wall is only 1/350 inch thick—thin but strong enough to resist deformation. The only threat to the wax structures is heat. On hot days, forager bees scatter tiny water droplets on the combs and fan their wings like crazy to prevent a meltdown.

The Uses of Beeswax Outside the Hive

The ancient Persians and Assyrians used beeswax to embalm their dead. In Egypt, cosmetologists added it to various facial creams, lip balms, nail polishes, and hair dressings. Artistic Egyptians mixed ground pigments into molten beeswax to make a special kind of paint called encaustic, ideal for portraying figures on flat slabs of stone. Among the Maya, Inca, Babylonians, and others, beeswax was used in the process of lost-wax casting, especially for fine silver and gold jewelry. Greek medical practitioners prescribed beeswax, liquefied and diluted with water, to soothe mucous membranes and curb dysentery and diarrhea. Medieval Europeans sealed important documents with beeswax to guard the intrigues they were forever plotting.

One of the most enduring human uses of beeswax is in the production of candles. Beeswax candles are famous for their pleasant fragrance and bright, clean-burning flame. Beeswax was probably one of the first waxes used in candlemaking and was far superior to the smoky, foul-smelling candles the poor made from rendered beef tallow. Today, paraffin, a by-product of petroleum refining, is used to make cheaper candles.

Probably the largest consumer of beeswax candles is the Catholic Church. In the past, candles made of beeswax secreted by "virgin bees" symbolized the virgin birth of Christ, his "spotless" body, and the pure light of the world. The wick represented Christ's soul, the flame his divine and human nature. The burning of sacramental candles reminded the faithful of Christ's death on the cross. Beeswax candles are still blessed by Catholic priests on Candlemas Day, celebrated on February 2. By papal decree, they must be at least 51 percent beeswax. Originally the Church required that sanctuary candles be 100 percent beeswax.

But candles are only part of the story. Today, there are literally hundreds of commercial products made with beeswax, ranging from cosmetics, skin ointments, stick colognes, and antiperspirants to shoe polishes, furniture and automobile waxes, industrial lubricants, and all kinds of anticorrosion coatings.

Thanks to all these uses, beeswax is big business. The Food and Agriculture Organization of the United Nations has estimated the world production of beeswax at forty-four million pounds a year, worth about $25 million.

Bee Pollen

Bees don't actually produce pollen; they just collect it, make it into bee bread, and store it in their pantries. Pollen grains are actually produced by the anthers of flowering plants. Inside the tough outer shell of each grain are the plants' sex cells, roughly the equivalent of sperm cells in animals. The sex cells are suspended in an inner goo, called cytoplasm, that is rich in proteins, nitrogen, nucleic acids, amino acids, lipids, antioxidants, and hundreds of other nutrients. This is what makes pollen the perfect bee food. Nectar and honey contribute energizing sugars to the bee diet and little else.

Each honey bee colony collects from twenty to forty pounds of pollen a year. Foragers add regurgitated nectar and saliva to the pollen grains they find on the flowers until they attain a Play-Doh-like consistency and can be easily transported in the "pollen baskets" on each of their hind legs. Back at the nest, a pollen-laden forager looks for an empty cell close to the outermost edge of brood on a brood comb. She then backs in and rubs her hind legs together to dislodge the two pollen pellets. Other foragers deposit their pellets until the cell is about half full. House bees then tamp and poke the pollen, adding honey, saliva, and other secretions so it will ripen into "bee bread."

Pollen for People?

Hyped by pollen prophets (profits?) and attractively packaged by entrepreneurs, bee pollen has been marketed as a miracle food for humans for decades. But the fact is, none of the health claims made for pollen have been substantiated in properly controlled clinical trials. Though it is high in proteins, lipids, antioxidants, and vitamins, these nutrients can be obtained in other, more easily digested foods at considerably less cost.

And then there are the side effects some people experience when taking pollen. The major adverse reactions are stomach pain and diarrhea, reported by up to 33 percent of individuals in some studies. Irritation or itching of the mouth and throat are also sometimes reported. So leave bee pollen to the bees, and enjoy their honey instead.

Propolis (Bee Glue)

Beekeepers may use Elmer's Glue-All to strengthen the joints in all those wooden Langstroth frames, but honey bees have a special glue all their own. Beekeepers call the brown, yellow, or green substance propolis and hate it because it sticks frames and hive parts together and makes their job more difficult.

Honey bees produce propolis from the resins, saps, and gums that they scrape from the stems and leaves of flowering plants, especially during the spring. Plants have a long history of producing these chemical compounds to use in their constant struggle with fungi, bacteria, and herbivorous animals and insects. Some of the chemicals are poison, like the glycosides present in milkweed sap, while others provide sticky physical barriers that stop predators in their tracks.

Bees like to have plenty of propolis on hand. They use it for caulking, sealing, strengthening, lining, and varnishing just about everything in the hive. When they construct their nests in rock cavities or depressions in the wild, they "paint" a ring of propolis around the colony to keep marauding ants away.

Propolis for People?

Propolis has several components that could make it an effective therapeutic agent for humans. It contains flavones pinocembrin, galangin, and caffeic and ferulic acids that are active against many bacteria and may be useful in healing wounds. It also contains quercetin, a flavone that has both antiviral and capillary-strengthening properties. Some of the flavonoids in propolis are capable of scavenging free radicals and protecting lipids and may even inhibit melanoma and carcinoma tumor cells.

Though propolis was used medicinally by the ancients, it had fallen out of favor until a few decades ago. Today, it is used for alleviating gum disease and is sometimes added to toothpastes. It does not seem to be toxic to humans, even when taken in large doses, though it can cause contact dermatitis in some people.

Unfortunately, there have been no clinical studies to test the medical benefits of propolis other than in oral hygiene. That's too bad, because it's an area that might be well worth investigating.

Royal Jelly

You may have seen small bottles of royal jelly glistening from the shelves of cosmetics departments and wondered what it is, what it does, and why it's so expensive. Well, royal jelly is a product of the hive, secreted by hypopharyngeal glands in the heads of young worker bees. All bee larvae are fed royal jelly for the first three days of their larval lives. From the fourth day on, the drone and worker larvae are fed pollen and honey instead, while larvae destined to become queens continue to dine solely on royal jelly. Thus, the development of a fertilized bee egg into a queen or worker is simply a matter of differing nutrition. Scientists still don't quite understand how this happens.

Royal jelly is 67 percent water, 11 percent sugar, 5 percent fatty acids, and 13 percent crude protein. It also contains cholesterol and other sterols as well as zinc, iron, copper, manganese, and high levels of the B vitamins.

A Lucrative Business

A number of health and cosmetic properties have been attributed to royal jelly over the years, but none have been backed up with controlled clinical studies. Other than some exotic fatty acids, there is nothing special or magical to be found in the chemical composition of royal jelly, and nothing you can't get from other, much less expensive foods. Claims of tissue repair, moisturizing, and other dermatological benefits need to be clinically tested in order to be credible.

We often hear that royal jelly can be sex-enhancing. It's a nice idea, but unfortunately its gonadotropic properties were disproved fifty years ago in trials with rats, who failed to get sexier on the royal diet. Royal jelly does contain testosterone, but the average human male already has a million times that amount circulating in his bloodstream.

Royal jelly is big business in Asia, where manufacturers do their best to hype the royal connection, the mystique of the queen bee, and its touted antiaging powers. Most of the world's supply comes from factories in China, where women meticulously extract minute amounts from queen cells housed in special brood frames. China produces more than eight hundred tons of royal jelly a year. While its benefits, even in cosmetics, are questionable, royal jelly, at $6 or more an ounce, commands regal prices. Once again, glamour wins out over scientific fact and reason.

Bee Venom

Bee venom is a liquid produced in the honey bee's poison gland and delivered through the aculeus, a hypodermic-needle-like stinger. Only females have stingers, leaving the unarmed males dependent on their sisters for protection. Millions of years ago, the stinger started out as a device for inserting eggs into plant stems or animal prey so the developing larvae could feed on their flesh. Later it took on a defensive role, effectively delivering venom to fend off attackers.

Contrary to popular belief, a honey bee that uses her stinger doesn't die immediately. It's true that she is partially eviscerated when the stinger is pulled from her body, but she may live for several hours. Beekeepers are very familiar with these "stung-out," kamikaze-like bees. They've lost their barbed weapon and chemical ammunition, but nature has given them the upper hand in psychological warfare against much larger enemies. People can be easily harassed by the relentless dive-bombing of harmless, stung-out bees, desperately swatting at them and often doing more damage to themselves than to the bees.

Venom is a bitter-tasting soup of proteins, peptides, and other biochemicals. Melittin, the main peptide, is the one that causes pain at the sting site. As a general rule, humans will survive if they receive six or fewer stings per pound of body weight. They have a 50 percent chance of survival at eight stings per pound and will probably die at ten stings per pound. Basically, this means that the average healthy adult male or female weighing 150 pounds can safely withstand up to nine hundred stings. Stings numbering fifteen hundred or more, however, can lead to death from liver and kidney failure.

If an individual is highly allergic to bee venom, however, the above figures are meaningless, for just one sting could send him or her into anaphylactic shock, requiring immediate medical attention by trained professionals. These people usually know who they are and carry an Epi-Pen or similar emergency kit containing injectable epinephrine and chewable antihistamine tablets.

Happily, less than 1 percent of the U.S. population is afflicted with insect sting allergy, which is responsible for about forty deaths a year, of which about seventeen can be attributed to honey bee stings. In addition, one person on average dies every year after receiving hundreds of Africanized honey bee stings. Africanized bees (*Apis mellifera adansonii*) have claimed between one thousand and two thousand lives in Latin

America since their accidental release near São Paulo, Brazil, in 1957. By way of comparison, only ten people have died in the United States from Africanized bee stings since the bees' arrival from Mexico in 1994. You are at far greater risk playing golf during a lightning storm.

There may be medical benefits attributable to bee venom. Some people actually solicit bee stings, placing bees against their skin and forcing them to sting. This may sound a bit masochistic, but the practice is called apitherapy, a sort of acupuncture with bees. No one is quite sure when or where it originated, but adherents believe it relieves the pain and reduces the disfigurement of rheumatoid arthritis, a crippling inflammatory disease of the joints. Justin Schmidt and other biomedical researchers believe that bee venom may in fact act as an anti-inflammatory agent. Certainly there is promise here, and additional clinical studies are warranted.

There is actually a small market for bee venom. It typically sells for about $90,000 a pound, or $6,000 an ounce, and is mainly purchased by allergists who use it to test their patients for bee sting allergy. It's also used for venom desensitization. This procedure requires that ever-increasing amounts of bee venom be injected into allergic patients at regular intervals over a period of a year or more. The desensitization builds up antibodies to the venom in the patient's body, thereby protecting the person from risks of anaphylactic shock.

But how is bee venom collected? In the past, it was obtained by dissecting the bees' poison glands one at a time, a laborious process. In 1963, the technology was improved with the invention of a venom-collecting device by Charles Mraz and Roger Morse. It consists of a wooden board with a membrane covered by evenly spaced metal wires. The apparatus is placed between two brood supers and the wires are electrified. This doesn't kill the bees, but it causes them to sting the membrane. When removed, the venom is dried and manually scraped off the tough membrane. It took Charles Mraz thirty years to produce 6.6 pounds of pure, dried honey bee venom. He must have been a very patient man.

How to Avoid Bee Attacks—and, Failing That, What to Do if You Get Stung

If you see lots of angry bees flying your way, remain calm. Walk to shelter, preferably inside a house or a vehicle with the windows rolled up. Do not move fast or flail your arms, for that will only further alarm and excite the bees. And do not jump into a body of water. The bees will still be there when you come up for air. If you are nowhere near shelter, place a piece of light fabric over your face and head, ideally something that you can see through. The bees will try to sting your face and head, so this is the area that needs to be protected. Ducking into the neck of your shirt and pulling it up is a good defense. Holding your breath keeps bees from stinging your mouth and nose, since exhaled carbon dioxide is one of the cues that stimulates their stinging response.

If you get stung on your hand, put the hand in your pocket. This may keep you from getting stung again. Each time a bee stings, it deposits a bit of alarm pheromone that smells like a ripe banana. (The odor comes from isopentyl acetate, which actually is one of the chemicals in ripe bananas.) The alarm pheromone literally tells other bees where to sting. By the way, never eat a ripe banana near a beehive.

Once you are out of danger, examine yourself for stings. If you see a small brown or black object in your skin, this is the stinger. Using a fingernail, the edge of a knife blade, or a credit card, carefully scrape it out. Do not pinch the sting site, as that will force the venom deeper into the wound. If you can get the sting out in under a minute, chances are you will prevent venom from entering your bloodstream. If possible, wash the sting site with soapy water.

Normally, the pain of a sting will subside after a few minutes. However, the sting site may start to itch a few hours later, and the itch could last for a day or two. If within the first thirty minutes of being stung, you begin sweating profusely, have difficulty breathing, or experience blurred vision, get medical assistance immediately. These are the first signs of anaphylactic shock.

Some people confuse large local site reactions following a bee or wasp sting with anaphylactic shock. The fact is, in certain individuals, even one sting can result in redness, swelling, and itching far from the original sting site. In those individuals, a sting on a finger or wrist might result in a red, swollen, and painful arm up to the elbow. This reaction is scary, perhaps, but is not anaphylaxis and is not life-threatening.

Bee Brood (Honey, I Ate the Larvae)

Bee brood, or larvae, is featured on dinner menus in many parts of the world. But don't be shocked—it's perfectly safe and not at all bad-tasting. As I learned as a graduate student at UC Davis, tender white bee larvae make a delicious quiche. Bee larvae are fed on pollen and honey and are probably much more wholesome than the hormone-laced beef, pork, poultry, and fish available in most stores. They are also higher in protein and lower in fat than beef and, unlike grasshoppers, crickets, and other edible insects, don't have an annoying, indigestible chitinous cuticle.

Brood as food actually has a long history. The earliest hunter-gatherers probably enjoyed larval bees straight from the comb. Modern honey hunters such as the Tongwe of Tanzania, the Shabanese of Congo, and the Kayapo of Brazil all quite happily chew combs containing larval and pupal honey bees, which provide an important part of their annual protein intake. In Asian markets, combs of bee brood are commonly sold alongside honeycombs and bottled honey.

In the Canadian provinces of Manitoba, Saskatchewan, and Alberta, it was common for bee colonies to be killed in the winter, at the time of honey extraction. The colonies were then started anew with packaged bees. This represented a huge waste, not only of bees but also of the protein in their larvae. Back in 1960 Brian Hocking and Fumio Matsumara estimated that 132 tons of bee brood were lost when the colonies were destroyed. They tried to develop a human market for bee brood, experimenting with both deep-fried larvae and baked larvae, cooked alone or in combination with other foods. Tasting panels preferred the deep-fried version, describing it as walnutlike in flavor and pork-crackling-like in texture. But the venture never got off the ground. People in the West are very squeamish about insects, and ingrained food habits are slow to change.

In Nepal, a bee brood dish called *bakuti* is a family favorite. It's made by squeezing brood combs through a woven mesh bag and collecting the milky juice. The liquid is gently heated for about five minutes while being constantly stirred. The result is similar in texture and flavor to scrambled eggs. In a taste test conducted in the United States, 85 percent of participants found bee brood à la Nepalese perfectly acceptable.

Bee larvae also make excellent pet food. Songbirds and reptiles are especially fond of it, and their owners would probably pay handsomely if supplies were regularly available. The beneficial-insects industry found that ladybird beetles and lacewings (beneficial insect predators that eat pests), normally difficult to raise in commercial insectaries, do very well when reared on diets containing bee larvae. If you are an entrepreneurial beekeeper looking for alternative markets, maybe the pet-food industry is right for you.

If fishing, is your sport, honey bee larvae and pupae make great bait. I owe this tip to a former co-worker, bee researcher Norbert Kauffeld, now living in Baton Rouge, Louisiana. How are they biting, Norb?

Appendix 4: The Chemical Composition of Honey

Component	Average	Range
Fructose/glucose ratio	1.23	0.76–1.86
Fructose (%)	38.38	30.91–44.26
Glucose (%)	30.31	22.89–40.75
Minerals (ash)	0.169	0.02–1.03
Moisture (%)	17.2	13.4–22.9
Reducing sugars (%)	76.75	61.39–83.72
Sucrose (%)	1.31	0.25–7.57
pH	3.91	
Total acidity (meq/kg)	29.12	8.68–59.49
True protein (mg/100g)	168.6	57.7–567

Table from the National Honey Board, www.nhb.org/foodtech/defdoc.html

Appendix 5: Resources

Mail Order Honeys

A. G. Ferrari
14234 Catalina Street
San Leandro, CA 94577
877.878.2783
www.agferrari.com
U.S. distributor of Sicilian and Tuscan honeys

Bosque Honey Farms
600 North Bosque Loop
Bosque Farms, NM 87068
505.869.2841
New Mexico raw honeys and bee pollen

Burt's Bees, Inc.
8221-A Brownleigh Drive
Raleigh, NC
www1.burtsbees.com

Candover Valley Honey Farm
2, Hackwood Cottages
Alton Road
Basingstoke, Hampshire
England, RG21 32 BA
01256.329064

Cannon Bee Honey Company
6105 11th Avenue South
Minneapolis, MN 55417
612.861.8999
Minnesota honeys including basswood, buckwheat, and clover

Cowboy Honey Company
P.O. Box 1387
Camp Verde, AZ 86322
520.567.3204
Arizona's famous mesquite honey

Dean & Deluca
2526 East 36th Street
Circle North
Wichita, KS 67219
877.826.9246
www.deandeluca.com
American, French, German, and Italian honeys

Derwent Valley Apiaries
RSD 1268 Lyell Highway
New Norfolk, Tasmania 7140
03.6261.1764
Tasmanian leatherwood honey

Fauchon
442 Park Avenue (at 56th Street)
New York, NY 10022
212.308.5919
www.fauchon.com
Rare honeys

Grossman Organic Farm
P.O. Box 1028
Tualatin, OR 97062
888.688.2582
Raw wildflower honeys and bee pollen from Oregon

Guilmette's Busy Bees
5539 Noon Road
Bellingham, WA 98226
360.398.2146
Honey from the Pacific Northwest, including fireweed, raspberry, and wildflower

Harold Curtis Honey Company
P.O. Box 1012
La Belle, FL 33975
Mangrove, orange blossom, and palmetto honeys

Hive Honey Shop
93 Northcote Road
London SW11 6PL
020.7924.6233
Various English honeys, honey condiments, beauty products made with honey, and tableware

Honey Garden Apiaries
P.O. Box 189
Hinesburg, VT 05461
802.985.5852
Raw wildflower honeys from New York State and Vermont, honey cough syrups, and beeswax candles

Hunter's Honey Farm
3440 Hancock Ridge Road
Martinsville, IN 46151
765.537.9430
Wildflower honeys from Indiana, beeswax candles, and pollen

Marshall's Farm
P.O. Box 10880
Napa, CA 94581
www.marshallshoney.com
Regional honeys from Napa Valley and San Francisco Bay area

McEvoy Ranch
P.O. Box 341
Petaluma, CA 94953
707.769.4122
Lavender honey from Sonoma County

National Honey Board
390 Lashley Street
Longmont, CO 80501
303.776.2337
www.nhb.org and www.honey.com
Sources for honey: www.honeylocator.com

Plan Bee Honey
17 Van Dam Street
New York, NY 10013
212.627.0046
www.planbeehoney.com
Wildflower honeys from New York State, comb honey, herb-infused honeys

River Hill Bee Farm
459 River Hill Road
Sparta, NC 28675
888.403.0392
www.riverhillhoney.com
Rare wildflower honeys, including sourwood from the Blue Ridge Mountains

Tropical Blossom Honey Company
106 N. Ridgewood Avenue
Edgewater, FL 32132
386.428.9027
www.tropicbeehoney.com
Orange blossom honey from Florida, other regional honeys, including palmetto and tupelo, flavored honeys

Uvalde Honey
P.O. Box 307
Uvalde, TX 78802
Honey from acacia blossoms (huajillo)

Craft and Hobby Supplies

GloryBee Foods, Inc.
P.O. Box 2744
Eugene, OR 97402
800.456.4923
www.glorybee.com
Beeswax candlemaking supplies, beekeeping equipment

Beekeeping Books

Avitable, A., and D. Sammataro. 1998. *The Beekeepers Handbook,* 3rd ed. Cornell University Press, Ithaca, NY.

Bonney, R. 1993. *Beekeeping: A Practical Guide.* Storey Books, Pownell, VT.

Bonney, R. 1991. *Hive Management: A Seasonal Guide for Beekeepers.* Storey Books, Pownell, VT.

Flottam, K. (ed.). 1988. *The New Starting Right with Bees,* 21st ed. A. I. Root Company, Medina, OH.

Graham, J. M. 1992. *The Hive and the Honey Bee.* Dadant and Sons, Hamilton, IL.

Longgood, W. 1985. *The Queen Must Die: And Other Affairs of Bees and Men.* W. W. Norton & Company, New York.

Morse, R. (ed.). 1990. *The ABC and XYZ of Bee Culture,* 40th ed. A. I. Root Company, Medina, OH.

Morse, R. 1986. *The Complete Guide to Beekeeping,* 3rd ed. E. P. Dutton, New York.

Morse, R. 1983. *A Year in the Beeyard.* Charles Scribner's Sons, New York.

Beekeeping Journals

American Bee Journal
Dadant and Sons, Inc.
51 South 2nd Street
Hamilton, IL 62341
800.637.7468
www.dadant.com

Bee Culture
A. I. Root Company
623 W. Liberty Street
Medina, OH 44256
800.289.7668
www.beeculture.com

Honey Processing and Beekeeping Equipment

To find local beekeeping clubs, try:
www.bee.airoot.com/beeculture/who.html.

3B Sales & Service
P.O. Box 6054
North Logan, UT 84341
(435) 258–2009
3bsales&service@pcu.net

A. I. Root Company
623 West Liberty Street
Medina, OH 44256
800.289.7668
www.rootcandles.com

Bee Maid Honey
625 Roseberry Street
Winnipeg, Manitoba
Canada R2H 0T4
204.783.2240

Better Bee, Inc.
8 Meader Road
Greenwich, NY 13834
800.632.3379

BHC Honey Suppliers
Unit 4, Wyeside Enterprise Park
Llanelwedd, Builth Wells
Wales, UK LD2 3UA
+44(0)1982 553235

Brushy Mountain Bee Farm
610 Bethany Church Road
Moravian Falls, NC 28654
800.233.7929
www.beeequipment.com

Dadant and Sons, Inc.
51 South 2nd Street
Hamilton, IL 62341
800.637.7468
www.dadant.com

Entomo-Logic
Evan Sugden and Kristina Williams
21323 232nd Street SE
Monroe, WA 98272–8982
360.863.8547
www.seanet.com/~entomologic/
entomologic_home.htmleasug
den@seanet.com

Garvin Honey Company
Avenue Three, Station Lane,
Witney, Oxon
United Kingdom OX8 6HZ
01993.775423

GloryBee Foods, Inc.
P.O. Box 2744
Eugene, OR 97402
800.456.4923
www.glorybee.com

Hector's Apiaries Services
2297 Stanislaws Court
Santa Rosa, CA 95401
707.579.9416

International Pollination, Inc.
International Pollination Systems
Dr. Ron Bitner
16645 Plum Lane
Caldwell, ID 83605
208.454.0086
www.pollination.com

Judi's Farm Market
8020 Steveston Highway
Richmond, British Columbia
Canada, V7A 1M3
604.275.9535

Kidd Bros. Produce Ltd.
5312 Grimmer Street
Burnaby, British Columbia
Canada, V5H 2H2
604.437.9757

Knox Cellars
Brian Griffin
1607 Knox Ave.
Bellingham, WA 98225
360.733.3283
www.knoxcellars.com
Sales@knoxcellars.com
Brian@knoxcellars.com

Mann Lake Ltd.
501 S. First Street
Hackensack, MN 56452
800.233.6663
www.mannlakeltd.com

Maxant Industries, Inc.
P.O. Box 453-S
Ayer, MA 01432
978.772.0576

Pollinator Paradise
Karen Strickler
31140 Circle Drive
Parma, ID 83660
www.pollinatorparadise.com
solitary@pollinatorparadise.com

Ruhl Bee Supply
12713 NE Whitaker Way
Portland, OR 97230
503.256.4231

Western Bee Supplies, Inc.
P.O. Box 190
Polson, MT 59860
800.548.8440
www.westernbee.com

Sources of Native Bees (Non-Apis) and Solitary Bee Nests

Custom Paper Tubes, Inc.
P.O. Box 44187
Cleveland, OH 44144–0187
216.362.2964
800.343.8823
www.custompapertubes.com
beetubes@custompapertubes.com

Sources

Introduction

Fellow entomologist Edward O. Wilson of Harvard University has influenced my thinking and writing perhaps more than anyone else. It has been my pleasure to join Ed on ant-collecting trips into the sky island mountaintop archipelagos of southern Arizona. Wilson also graced *The Forgotten Pollinators,* the book I wrote with Gary Paul Nabhan, with a prescient foreword explaining why pollination matters to the human race. Wilson's recent books on biophilia and conservation are must-reads for anyone interested in the natural world and our complex relationship with it.

Kellert, S. R., and E. O. Wilson (eds.). 1993. *The Biophilia Hypothesis.* Island Press, Washington, DC.

Kellert, S. R. 1997. *Kinship to Mastery: Biophilia in Human Evolution and Development.* Island Press, Washington, DC.

Pyle, R. M. 1993. *The Thunder Tree: Lessons from an Urban Wildland.* Houghton Mifflin Co., Boston, MA.

Wilson, E. O. 1971. *The Insect Societies.* Belknap Press, Cambridge, MA.

Wilson, E. O. 1975. *Sociobiology: The New Synthesis.* Belknap Press, Cambridge, MA.

Wilson, E. O. 1984. *Biophilia: The Human Bond with Other Species.* Harvard University Press, Cambridge, MA.

Wilson, E. O. 1992. *The Diversity of Life.* Belknap Press, Cambridge, MA.

Wilson, E. O. 1996. *In Search of Nature.* Island Press, Washington, DC.

Wilson, E. O. 2002. *The Future of Life.* Alfred A. Knopf, New York.

Chapter 1. The Beginning of an Enduring Passion: Prehistoric Honey Hunters

The books by Dr. Eva Crane, former director of the International Bee Research Association in Cardiff, Wales, are treasure troves of information on prehistoric petroglyphs depicting ancient honey hunts. We thank Eva for allowing us to reproduce some of the evocative illustrations found in these marvelous works. The 1937 book by Hilda Ransome, unfortunately out of print, contains facts about bees and honey in the ancient world that you won't find anywhere else.

Crane, E. 1983. *The Archaeology of Beekeeping.* International Bee Research Association, Cardiff, Wales.

Crane, E. 2001. *The Rock Art of Honey Hunters.* International Bee Research Association. Cardiff, Wales.

Ransome, H. M. 1937. *The Sacred Bee in Ancient Times and Folklore.* George Allen & Unwin, London.

Chapter 2. Searching for Gold: Ancient Rituals and Modern-Day Honey Hunters

I was particularly moved by the words and images of Eric Valli and Diane Summers as they recorded, both in print and on film in a National Geographic television special, the adventures of Gurung honey hunters in the foothills of the Himalayas. The color photographs in their elegant book shouldn't be missed.

Buchmann, S. L., and G. P. Nabhan. 1997. *The Forgotten Pollinators.* Island Press, Washington, DC.

Crane, E. 1983. *The Archaeology of Beekeeping.* Gerald Duckworth, London.

Dollin, L. 2001. "Australian Native Bees—Treasured in Aboriginal Heritage, Part 2." *Aussie Bee* 16: 15–17. Australian Native Bee Research Centre, North Richmond NSW, Australia.

Dollin, L. 2000. "Australian Native Bees—Treasured in Aboriginal Heritage, Part 1." *Aussie Bee* 15: 10–12. Australian Native Bee Research Centre, North Richmond NSW, Australia.

Dollin, L. 2000. "Overland Expedition for an Elusive Drone." *Aussie Bee* 15: 8–9. Australian Native Bee Research Centre, North Richmond NSW, Australia.

Dollin, L. 1997. "Exploring Western Australia—Part 2: An Old Native Bee Farming Area of the Aborigines." *Aussie Bee* 4: 14–15. Australian Native Bee Research Centre, North Richmond NSW, Australia.

Ransome, H. M. 1937. *The Sacred Bee in Ancient Times and Folklore.* George Allen & Unwin, London.

Valli, E., and D. Summers. 1988. *Honey Hunters of Nepal.* Harry N. Abrams, New York.

Chapter 3. Staying in Touch: The Beekeeper's Craft

The books by Dr. Eva Crane and Hilda Ransome are good introductions to the history of beekeeping. The articles by Donald Brand and the Weavers will take you into the fascinating world of meliponiculture, the keeping of stingless bees by the ancient and modern Maya.

Berenbaum, M. R. 1995. *Bugs in the System: Insects and Their Impact on Human Affairs.* Addison-Wesley, Reading, MA.

Brand, D. D. 1988. "The Honey Bee in New Spain and Mexico." *Journal of Cultural Geography* 9, no. 1: 71–82.

Crane, E. 1992. "The Past and Present Status of Beekeeping with Stingless Bees." *Bee World* 73, no. 1: 29–42.

Crane, E. 1990. *Bees and Beekeeping: Science, Practice and World Resources.* Cornell University Press, Ithaca, NY.

Crane, E. 1983. *The Archaeology of Beekeeping.* Gerald Duckworth, London.

Ransome, H. M. 1937. *The Sacred Bee in Ancient Times and Folklore.* George Allen & Unwin, London.

Virgil. 1982. *Georgics,* trans. L. P. Wilkinson. Penguin Books, New York.

Weaver, N., and E. C. Weaver. 1981. "Beekeeping with the stingless bee, Melipona beecheii, by the Yucatecan Maya." *Bee World* 62, no. 1: 7–19.

Chapter 4. A Year in the Life of a Beekeeper

Elegant prose describing the life and seasons of the beekeeper can be found in the books by Sue Hubbell and Douglas Whynott, and in *The Queen Must Die,* an often overlooked but essential work by the late William Longgood. I also recommend *The Hive and the Honey Bee* and *The ABC and XYZ of Bee Culture.*

Crane, E. 1990. *Bees and Beekeeping: Science, Practice and World Resources.* Cornell University Press, Ithaca, NY.

Graham, J. (ed.). 1992. *The Hive and the Honey Bee.* Dadant and Sons, Hamilton, IL.

Hubbell, S. 1988. *A Book of Bees.* Random House, New York.

Hubbell, S. 1986. *A Country Year.* Random House, New York.

Kelly, K. 1995. *Out of Control: The New Biology of Machines, Social Systems, and the Economic World.* Addison-Wesley, Reading, MA.

Longgood, W. 1985. *The Queen Must Die: And Other Affairs of Bees and Men.* W. W. Norton, New York.

Morse, R. A. (ed.). 1990. *The ABC and XYZ of Bee Culture,* 40th ed. A. I. Root, Medina, OH.

Morse, R. A. 1986. *The Complete Guide to Beekeeping,* 3rd ed. E. P. Dutton, New York.

Morse, R. A., and T. Hooper (eds.). 1985. *The Illustrated Encyclopedia of Beekeeping.* Alphabooks, Sherborne, UK.

Whynott, D. 1991. *Following the Bloom: Across America with the Migratory Beekeepers.* Stackpole Books, Harrisburg, PA.

Chapter 5. Secrets of the Bee

The Sex Life of Flowers, by Meeuse and Morris, plus the PBS special of the same name, is an excellent introduction to the sexy world of pollinators.

Bernhardt, P. 1999. *The Rose's Kiss: A Natural History of Flowers.* Island Press, Washington, DC.

Bernhardt, P. 1993. *Natural Affairs: A Botanist Looks at the Attachments Between Plants and People.* Villard, New York.

Darwin, C. 1890. *The Various Contrivances by Which Orchids Are Fertilized by Insects,* 2nd ed. Murray, London.

Frisch, K. von. 1967. *The Dance Language and Orientation of Bees.* Harvard University Press, Cambridge, MA.

Gould, J. L., and C. G. Gould. 1995. *The Honey Bee.* W. H. Freeman, New York.

Graham, J. (ed.). 1992. *The Hive and the Honey Bee.* Dadant and Sons, Hamilton, IL.

Heinrich, B. 1979. *Bumble-Bee Economics.* Harvard University Press, Cambridge, MA.

Maeterlinck, M. 1954. *The Life of the Bee.* Dodd, Mead, New York.

Meeuse, B., and S. Morris. 1984. *The Sex Life of Flowers.* Facts on File, New York.

Michener, C. D. 2000. *Bees of the World.* The Johns Hopkins University Press, Baltimore, MD.

Michener, C. D. 1974. *The Social Behavior of the Bees: A Comparative Study.* Belknap Press, Cambridge, MA.

Morse, R. A. (ed.). 1990. *The ABC and XYZ of Bee Culture,* 40th ed. A. I. Root, Medina, OH.

Roubik, D. W. 1989. *Ecology and Natural History of Tropical Bees.* Cambridge University Press, Cambridge, UK.

Seeley, T. D. 1995. *The Wisdom of the Hive: The Social Physiology of Honey Bee Colonies.* Harvard University Press, Cambridge, MA.

Seeley, T. D. 1985. *Honey Bee Ecology: A Study of Adaptation in Social Life.* Princeton University Press, Princeton, NJ.

Chapter 6. Bees and Honey in Myth, Legend, and Ancient Warfare

The writings of Dr. Eva Crane provide a rich synthesis of archaeological findings related to the role of honey bees in myths, legends, and ancient battles.

Crane, E. 1980. *A Book of Honey.* Oxford University Press, Oxford, UK.

Crane, E. (ed.) 1975. *Honey: A Comprehensive Survey.* Crane Russak, New York.
Ransome, H. M. 1937. *The Sacred Bee in Ancient Times and Folklore.* George Allen & Unwin, London.
Reed, A. W. 1965. *Aboriginal Fables and Legendary Tales.* A. H. & A. W. Reed, Sydney, Australia.

Chapter 7. Trading Honey in the Ancient and Modern Worlds

Good sources on the honey trade include, once again, the works of Dr. Eva Crane and Donald Brand.

Brand, D. D. 1988. "The Honey Bee in New Spain and Mexico." *Journal of Cultural Geography* 9, no. 1: 71–82.
Crane, E. 1980. *A Book of Honey.* Oxford University Press, Oxford, UK.
Crane, E. (ed.). 1975. *Honey: A Comprehensive Survey.* Crane, Russak, New York.
Toussaint-Samat, M. 1993. *History of Food.* Blackwell, Oxford, UK.
Virgil. 1982. *Georgics,* trans. L. P. Wilkinson. Penguin Books, New York.
Virgil. 1965. *Eclogues and Georgics.* J. M. Dent and Sons, London.

Chapter 8. A Taste of Honey: Sampling Varieties from Around the World

The recent books by Stephanie Rosenbaum and Sue Style provide an excellent entree into the world of boutique honeys, monofloral sources, and exciting varietals. The Web site of the National Honey Board is also a great place to learn about different honeys and where to buy them. (See Appendix 5 for other sources of both common and rare honeys.)

Crane, E. (ed.). 1975. *Honey: A Comprehensive Survey.* Crane, Russak, New York.
Rosenbaum, S. 2002. *Honey: From Flower to Table.* Chronicle Books, San Francisco.
Style, S. 1993. *Honey: From Hive to Honeypot.* Chronicle Books, San Francisco.

Chapter 9. How Sweet It Is: Cooking with Honey Throughout the Ages

One of the best histories of cooking with honey is Maguelonne Toussaint-Samat's *History of Food.* Harold McGee's book is an entertaining and informative account of cooking with honey today. But don't just read about cooking with honey; try some recipes in this book.

Brothwell, D., and P. Brothwell. 1969. *Food in Antiquity.* Frederick A. Praeger, New York.

Corriher, S. 1997. *Cookwise.* William Morrow, New York.
McGee, H. 1984. *On Food and Cooking.* Scribner, New York.
Roden, C. 1996. *A Book of Jewish Food.* Alfred A. Knopf, New York.
Rosenbaum, S. 2002. *Honey: From Flower to Table.* Chronicle Books, San Francisco.
Style, S. 1993. *Honey: From Hive to Honeypot.* Chronicle Books, San Francisco.
Tannahill, R. 1989. *Food in History.* Crown, New York.
Toussaint-Samat, M. 1993. *History of Food.* Blackwell, Oxford, UK.
Witty, H. 1997. *The Good Stuff Cookbook.* Workman, New York.

Chapter 10. Mead: The Honey that Goes to Your Head

Readers wishing to try their hand at the ancient art of mead making should refer to the books by Acton and Duncan and by the late Roger Morse of Cornell University. Newcomers to the world of mead will find that some are dry and light while others are the very sweet varieties that most people think of. Unfortunately, there are very few meads available commercially in the United States.

Acton, B., and P. Duncan. 1984. *Making Mead: A Complete Guide to the Making of Sweet and Dry Mead, Melomel, Metheglin, Hippocras, Pyment and Cyser.* G. W. Kent, Ann Arbor, MI.
Crane, E. (ed.). 1975. *Honey: A Comprehensive Survey.* Crane, Russak, New York.
Gayre, G. R. 1948. *Wassail! In the Mazers of Mead.* Phillimore, London
Morse, R. A. 1992. *Making Mead Honey Wine: History, Recipes, Methods and Equipment.* Wicwas Press, Ithaca, New York.
Renfrow, C. 1994. *A Sip Through Time: A Collection of Old Brewing Recipes.* C. Renfrow, n.p.
Spence, P. 1997. *Mad About Mead! Nectar of the Gods.* Llewellyn Publications, St. Paul, MN.

Chapter 11. Good for What Ails You

I'm thrilled to see this "antiquated" folk medicine being reevaluated and clinically tested by modern medical researchers. The pioneering work by Dr. Peter Molan and his colleagues with manuka honey is particularly promising.

Armstrong, S., and G. W. Otis. 1995. "The Antibacterial Properties of Honey." *Bee Culture* 123, no. 9: 500–2.
Beck, B. 1938. *Honey and Health.* Robert McBride, New York.
Breaste, J. H. 1930. *The Edwin Smith Surgical Papyrus.* University of Chicago Press, Chicago.
Guido, M. 1975. *The Healing Hand: Man and Wound in the Ancient World.* Harvard University Press, Cambridge, MA.

Han, H., G. E. Miller, and N. DeVille. 2003. *Ancient Herbs, Modern Medicine.* Bantam, New York.

Jarvis, D. C. 1958. *Folk Medicine.* Holt, Rinehart and Winston, New York.

Molan, P. C. 1999. "Why Honey Is Effective as a Medicine. 1. Its Use in Modern Medicine." *Bee World* 80, no. 2: 80–92.

Molan, P. C. 1999. "Potential of Honey in the Treatment of Wounds and Burns." *American Journal of Clinical Dermatology* 2, no. 1: 13–19.

Molan, P. C. 1998. "A Brief Review of Honey as a Clinical Dressing." *Primary Intention* 6, no. 4: 148–58.

Molan, P. C. 1992. "The Antibacterial Activity of Honey. 1. The Nature of the Antibacterial Activity." *Bee World* 73, no. 1: 528.

Molan, P. C. 1992. "The Antibacterial Activity of Honey. 2. Variation in the Potency of the Antibacterial Activity." *Bee World* 73, no. 2: 59–76.

Munn, P., and R. Jones (eds.). 2001. *Honey and Healing.* International Bee Research Association, Cardiff, Wales.

Traynor, J. 2002. *Honey: The Gourmet Medicine.* Kovak Books, Bakersfield, CA.

Permissions

Illustration Credits

Pages 12,14,17: Crane, Eva, ed. *The Rock Art of Honey Hunters*. Cardiff, UK: International Bee Research Association, 2001.

Pages 23, 71: Stephen Buchmann

Pages 33, 66: Paul Mirocha

Page 40: Viv Sinnamon

Page 48: Tory Sagrillo

Pages 52, 53: Crane, Eva. *The Archaeology of Beekeeping*. Ithaca, NY: Cornell University Press, 1983. Courtesy of Gerald Duckwork & Co.

Pages 56: Museo de America, Madrid

Page 59: Anton, Ferdinand. *Art of the Maya*. London: Thames and Hudson, 1970.

Page 76: Division of Agriculture and Natural Resources, University of California, Davis.

Pages 97, 106, 112: Scott Camazine

Page 137: Jones, Proctor Patterson. *Napoleon: An Intimate Account of the Years of Supremacy, 1800–1814*. San Francisco, CA: Proctor Jones Publishing Co., 1992.

Page 200: Lester, G. A. *The Anglo-Saxons: How They Lived and Worked*. Newton Abbot, UK: David & Charles, 1976.

Recipe Permissions

Pages 172: Yash Senturk, North Cyprus Home Page

Pages 179: Mica at www.myhouseandgarden.com

Pages 181, 183, 186–190: National Honey Board

Page 191–192: The Kitchen of Stephen Buchmann

Index

Page numbers of illustrations appear in italics.

About the Authors

STEPHEN BUCHMANN is an amateur beekeeper, associate professor of entomology at the University of Arizona in Tucson, co-author of *The Forgotten Pollinators,* and founder of The Bee Works, an environmental company.

BANNING REPPLIER is a writer who lives in New York City.